A NATURALIST'S GUIDE TO THE

BIRDS
OF
INDIA

Bangladesh, Bhutan, Nepal, Pakistan and Sri Lanka

Bikram Grewal and Garima Bhatia

JOHN BEAUFOY PUBLISHING

This edition published in the United Kingdom in 2022 by John Beaufoy Publishing
11 Blenheim Court, 316 Woodstock Road, Oxford OX2 7NS, England
www.johnbeaufoy.com

10 9 8 7 6 5 4 3 2 1

Photo Credits
Front cover: *main image* Spot-bellied Eagle Owl (Varun Thakkar), *bottom left* Black and Orange Flycatcher (Anand Israel), *bottom centre* Hill Myna (Clement M Francis), *bottom right* Oriental Dwarf Kingfisher (Manjula Mathur).
Back cover: Scaly-breasted Munia (Garima Bhatia). **Title page:** Indian Pitta (Biswaroop Satpati). **Contents page:** Scarlet Minivet (Panchami Mano Ukil)
Main descriptions: photos are denoted by a page number followed by T (top), B (bottom), L (left), R (right), a (above), b (below).

Abhay kewat: 55Ta; Abhinash Dhal: 67TR; **Amano Samarpan:** 73TL; **Amit Sharma:** 39T, 39BR, 48T, 58BR, 87BL, 114B, 145BL; **Amit Thakurta:** 81BL; **Anand Gupta:** 80TL, 108B; **Arpit Deomurari:** 21B, 41Tb, 44BR, 55Tb, 114T, 120TR, 126TL; **Ashok Kayshap:** 28B, 31TL, 61B; **Atanu Bose:** 98BR; **Avinash Khemka:** 40Ta, 107B; **Bhanu Singh:** 39BL, 122BL; **Bishan Monappa:** 54BR, 68TR, 70BR, 113Tb, 133B; **Biswapriya Rahut:** 147BL; **Biswarup Satpati:** 13Tb, 24T, 31B, 37BL, 38T, 62BR, 63T, 64B, 67TL, 68TL, 70T, 72B, 75BR, 75BL, 78B, 79B, 80BL, 81T, 82TR, 82BL, 88TL, 89T, 90B, 110T, 112T, 115BR, 119BL, 123BR, 124B, 125T, 128TL, 128TR, 128Ba, 131BL, 132TR, 133Tbl, 136B, 143T; **Clement M Francis:** 23TL, 23B, 24BL, 25B, 26TR, 29BR, 30BL, 33B, 34BL, 34BR, 38B, 40BL, 45B, 87Ta, 95T, 96TL, 100BL, 117T, 121Ta, 127BL, 145Ta, 145Tb, 150B; **Deboshree Gogoi:** 91TL; **Dhritman Hore:** 138Tb, 149Tb, 149BL; **Dolly L Bhardwaj:** 83Bb, 130TR; **Garima Bhatia:** 12TR, 17BR, 19B, 20Ta, 20B, 22T, 24BR, 30T, 30BR, 32B, 35B, 41Ba, 43T, 45Tb, 47BL, 54BL, 58BL, 59TL, 73TR, 73TC, 77TL, 79TL, 82TL, 82BR, 84TL, 84TR, 85TR, 86T, 89B, 92B, 96TR, 98BL, 99TR, 99B, 100BR, 103BL, 106B, 110B, 111TL, 111BL, 115T, 116T, 116B, 117Ba, 118T, 118BL, 118BC, 122BR, 123BL, 126TR, 127BC, 127T, 128Bb, 129BL, 130BR, 138BR, 141B, 142T, 143BR, 148Ta, 148BR, 149Ta, 150TL; **Gopinath Kollur:** 35T, 56TL; **Gunjan Arora:** 26TL; **Gururaj Moorching:** 43B, 66TL, 112B; **Jainy Maria Kurikose:** 135TL; **Khushboo Sharma:** 57TL, 120TL, 131T, 146B; **Kintoo Dhawan:** 12TL, 56BR, 57B, 64T, 65B, 135TR, 143BL, 147T; **K Hari Krishnan:** 129TR; **Kshounish Shankar Roy:** 144TR; **Kunan Naik:** 16TR, 17TR, 21T, 51T, 78T, 94TL, 101B, 125B; **Manjula Mathur:** 14TR, 16TL, 17TL, 18BL, 22Bb, 23TR, 25T, 28T, 32TR, 36TL, 36TR, 37Ta, 40Tb, 41Bb, 47T, 47BR, 48B, 50B, 52T, 63BL, 66BR, 81BR, 83TR, 85BL, 86B, 90TR, 93T, 94B, 105T, 106T, 107T, 111BR, 119TR, 124T, 132TL, 133TR, 134BL, 139B, 140TR, 140B, 145BC, 148TL, 150TR, 151TR, 151TL; **Manoj Kejriwal:** 60TL, 79TL, 83TL, 84B, 85BR, 90TL, 100T, 102Ba, 108T, 123TL, 138Ta; **Megh Roy Choudhury:** 117Bb, 149BR; **Mousumi Dutta:** 95BR; **Nikhil Devasar:** 15Tb, 16BR, 18TL, 27B, 46BL, 51BL, 85TL, 126BR, 138BL, 146TL, 147BR, 148Tb; **Ninaad Kulkarni:** 137Ba, 144B; **Niranjan Sant:** 80BR, 104B; **Nitin Srinivasamurthy:** 18TC, 19T, 22Ba, 27TL, 29T, 29BL, 36B, 53B, 59BR, 75TR, 75TL, 91BL, 96BR, 98TR, 99TL, 122Tb, 126BL, 129TL, 129BR, 140TL, 142BL, 148BL; **Panchami Manoo Ukil:** 15B, 16BL, 27TR, 34T, 42TR, 42B, 42TR, 46T, 62BL, 72T, 73B, 88B, 102T, 120B, 121T, 121Bb, 130TL, 131BR, 139TR; **Pranesh Kodancha:** 111TR; **Prasad Basavraj:** 101T; **Rajat Bhargava:** 23TC; **Rajneesh Suvarna:** 56TR; **Ramki Sreenivasan:** 69TR, 91BR, 94TR, 95BL, 96BL, 122Ta, 131BC, 145BR; **Ranjan Barthakur:** 97T, 105B, 137T; **Samyak Kaninde:** 123TR; **SarwanDeep Singh:** 40BR, 49B, 57TR, 98BC, 137BL, 144TL; **Satisha Sarakki:** 80BC, 88TR; **Savio Fonseca:** 13B, 14TL, 14B, 15a, 17BL, 26B, 37BR, 41Ta, 46BR, 49T, 50T, 54T, 55B, 58T, 59TR, 60TR, 60B, 61T, 62T, 63BR, 65TL, 65TR, 66TR, 68B, 69TL, 70BL, 71T, 74TL, 74TR, 74B, 76T, 76B, 77TR, 87Tbl, 87BR, 97B, 102Bb, 103T, 109B, 113B, 115BL, 132B, 134T, 134BR, 135B, 136T, 137Bb, 146TR; **Shiv Sankar:** 83Ba; **Sujan Chatterjee:** 59BL; **Sumit K Sen:** 18TR, 33TR, 42TL, 52B, 80TR, 103BL; **Supriya Das:** 69B, 92T; **Supriyo Samanta:** 20Tb, 93B; **Tanmoy Das:** 31TR; **Vaidehi Gunjal:** 87Tbr, 104T, 141T; **Vinit Arora:** 37TB, 44T, 44BL, 56BL, 66BL, 67B, 113Ta, 133Ta, 151B

ISBN 978-1-913679-34-7

Edited, designed and typeset by Alpana Khare Graphic Design, New Delhi, India

Printed and bound in Malaysia by Times Offset (M) Sdn. Bhd.

·Contents·

INTRODUCTION

The Indian region is incredibly rich in birdlife. Over 1,300 of the world's species of bird are found in the region. This number rises to over 2,000 with subspecies included, which makes the Indian checklist twice the size of those of Europe and North America. This abundance is due to the variety of habitats and climate. Altitude ranges from sea level to the peaks of the Himalaya, the world's highest mountain range; rainfall from its lowest in the Rajasthan desert to its highest in the north-eastern town of Cherrapunji in Meghalaya, one of the wettest places in the world. Unlike in more temperate zones, the climate of large areas of the Indian region encourages continuous plant growth and insect activity – abundant sources of avian nourishment throughout the year.

ORIENTAL AND SUBREGIONS

In the 19th century, P. L. Sclater studied the world's birds and divided the planet into six biogeographic realms. This was later slightly modified to apply to all animals. The Oriental realm covers South and Southeast Asia, the Himalaya separating it from the Palearctic to the north. Leaf birds are found exclusively in this region and most broadbills too. In general the region's birds have closest affinities with those of tropical Africa.

The Indian subregion has further been divided into seven different areas, in which different types of bird are found.

The northernmost of these areas is the Himalaya, which forms an arc some 2,500km long and 150 to 400km broad across the top of the subcontinent. The Himalayan mountains form roughly three parts, the foothills or Sivaliks to the south, the Himachal, or lower mountains, and the Himadri or high Himalaya to the north. The Ladakh plateau, with an average elevation of 5,300m, occupies a large portion of the Indian state of Jammu and Kashmir and consists of steppe country with mountain lakes where birds like the Bar-headed Goose and Brown-headed Gull breed in summer. The state of Himachal Pradesh, and the Kumaon and Garhwal regions of the state of Uttarakhand, lie to the west of Nepal, which falls almost entirely within the central Himalaya. Further east the rainfall increases giving the Eastern Himalaya of Bhutan and Sikkim a very different range of species from those in the west.

The north-west covers the bulk of Pakistan, the flat plains of the Indian Punjab and the semi-arid and arid plains of Rajasthan in the west. The Punjab (divided now between India and Pakistan) is watered by the five rivers, after which it takes its name, and efficient farming on fertile soil means that it produces an immense surplus of wheat and rice. Further west, wherever irrigation has been possible the desert has bloomed. Mountains of red chillies, for example, can be seen drying next to the fields around Jodhpur, while there are verdant paddy fields in areas irrigated by the great River Indus in Pakistan's Sindh province. Areas without irrigation have to rely on the perennially deficient rainfall, but local grasses have adapted to this, and after a monsoon shower even the desert sprouts rich pasture. Much of the area is, in fact, thorn scrub rather than true desert. Among the numerous desert birds found in this area are many which are related to species further west. The shifting sands of the desert join ultimately with the

Rann of Kutch, a large salt waste which runs into the sea, and are bordered to the south-east by the Aravallis, India's most ancient mountains.

North India comprises the Gangetic plain, enriched by thousands of years of alluvial deposits brought by the River Ganga and her tributaries from the Himalaya. The Gangetic plain is densely populated and highly fertile. This region extends up to an altitude of 1,000m in the north, so that it includes the low foothills of the Himalaya, and the terai of India and Nepal, once a marshy area covered with dense forest. Much of the terai area has been cleared for farming but some of the forests which still exist reveal the fantastic variety of birdlife which these forests must once have supported.

Peninsular India, bordered on the north-west by the Aravallis and the north by the Vindhya mountains, on the west by the Arabian Sea and the east by the Bay of Bengal, makes up the largest physiographic division of India. The central plateaus of this area, which is also known as the Deccan, rise to over 1,000m in the south, but hardly exceed 500m in the north. The peninsula has some wonderful landscapes, hills and huge boulders littering the countryside, and large areas of forest. Great rivers like the Narmada rise in the heart of the peninsula and flow into the sea. The steep escarpments of the Western Ghats, the mountains which stand between the plateau and the low-lying coastal strip, catch the full force of the monsoon.

The south-west region lies within the peninsula, but due to the particularly humid climate and the height of the hills here, its birds, like spiderhunters and laughingthrushes, bear strong affinities with those found in the north-east and Myanmar. The highest of the hills here are the Nilgiris or Blue Mountains, much of whose characteristic downland and *shola* forest is now under eucalyptus, tea and other plantation crops. Tea is also the main crop of the Annamalai or Elephant Mountains of Kerala, while cardamom and other spices are grown lower down. Perhaps the most ornithologically fascinating part of this area are the forests of the Wayanad, where Kerala, Karnataka and Tamil Nadu meet.

The north-east and Bangladesh region consists of the delta of the Ganga and Brahmaputra, with its tidal estuaries, sandbanks, mudflats, mangrove swamps and islands. Further upstream are lands drained, and occasionally flooded, by these great rivers and their tributaries. The north-east region also extends northwards to include all the forest regions of the states of Arunachal Pradesh, Mizoram, Meghalaya and Nagaland, as well as the Kingdom of Bhutan. As you progress eastwards, the birdlife has increasingly strong affinities with the Indo-Chinese subregion.

Sri Lanka is a remarkable area for birdlife. Although far from large, the country has a wide range of climate and habitat which supports some 400 species and subspecies of bird, including 21 species like the Ceylon Blue Magpie found nowhere else. Many of the island's birds are identical to those found in India, although for some reason vultures have not been able to cross the Palk Strait. Sri Lanka can be divided into three zones, the dry plains of the north, the mountainous central region and the humid wet zone around the capital Colombo. The most useful detailed work exclusively dealing with Sri Lankan ornithology is G. M. Henry's A *Guide to the Birds of Ceylon.*

HABITAT

While many common species are spread over large areas of the Oriental realm, others are limited not just to a region but also to habitat. Some birds of the conifer forests of the hills will be found only there, while grassland birds may be restricted to that habitat.

As the subcontinent has a very dense human population, birds which get on well with man flourish. These are not limited to House Sparrows, crows and House Martins. Indian culture has traditionally respected all forms of life and protected birds before sanctuaries and parks were ever thought of. India's only resident crane, the Sarus, is left unharmed no matter how much of a farmer's pea crop it consumes. The Indian Peafowl has in areas a semi-sacred status, which is why it is found in large numbers undisturbed. The Red Junglefowl has a long history of association with man and is the ancestor of the domestic chicken.

City gardens are homes for many species, including tailorbirds, sunbirds, white-eyes, babblers and the ubiquitous myna. Other birds take advantage of cultivation techniques; especially Indian Pond Herons, who often take up position in paddy fields, practising what villagers call *bagla bhakti*–supreme hypocrisy: sitting like a holy man lost in meditation, but in fact just waiting to stab something in the back. Little Grebes also take up residence in village ponds, while refuse is in great demand by pariah kites which, along with pariah dogs, haunt the rubbish heaps of the subcontinent.

Shallow lagoons, inland *jheels* or shallow lakes, and rivers are rich habitats for waterbirds from pelicans, storks, cranes, egrets and cormorants to the jacanas and gallinules among the lotuses, and the waders picking their way along the water's edge probing for food. Huge numbers of migratory waterfowl also congregate at jheels during the winter months. Other birds, like bitterns, conceal themselves among reed beds. Numerous birds of prey can be found near water. The attractive Brahminy Kite is particularly adaptable. It can be seen from the lakes of north India to the sea coasts of the south. Many birds of the coast are distinctive. Typical is the Western Reef Egret, seldom found far inland, which feeds on molluscs and crustaceans.

In the more arid inland areas are found larks, chats, sandgrouse and the rare Great Indian Bustard, a stunning bird seen in small flocks which fly in to land like great avian aircraft. Desert birds tend to be sandy in colouration, which helps to camouflage them.

Forest birds are more difficult to spot, especially when they are concealed in the tree canopy. However, they can often be seen in clearings flying from tree to tree or on the edges of forests where the sun can penetrate and there is a great deal of insect activity. Often assorted species form hunting parties and move together through the forest. So in one place you can see woodpeckers, warblers, tits and treecreepers. Often birds can be located and identified through their calls. Here it is also worth noting that Oriental forests are home to many more birds than those nearer either pole. A tropical forest can hold more than 200 bird species at more than 5,000 pairs per km, but a northern forest may hold fewer than 20 species at 200 individuals per km.

ADAPTATION

The most outstanding visual signs of the way in which birds have adapted are of course, the conversion of their upper limbs into wings, the growth of feathers and the ability to fly.

Plumage is further adapted depending on habitat and habits. The feathers of cormorants and snakebirds get drenched to allow them to swim under water, but other birds effectively coat their feathers with oil from oil glands to waterproof them. Plumage and body shape are adapted for specialized flight. Owl feathers are so formed as to give silent flight. Built especially for speed is the Peregrine Falcon, which swoops on its prey from a great height. Colouring is also an important adaptation. Camouflage is seen, for example, in sandgrouse, snipe, bitterns, owls and nightjars. Teeth would weigh down the head of any bird wanting to fly efficiently, so over the past 100 million years birds have lost them and instead developed gizzards. The gizzard is situated near a bird's centre of gravity. Birds gulp food down into their crop from where it is ground down in the muscular gizzard with the aid of grit and small stones that the birds swallow. The ability to disgorge indigestible bones, fur, insect shells or large seeds in the form of pellets is another form of adaptation.

Birds have developed specialized bills and feet for feeding. The most generalized bill perhaps belongs to the omnivorous crow. It is straight, pointed and roughly triangular in section. Birds like herons and kingfishers have more dagger-like bills, suitable for catching fish and frogs. Not all kingfishers need water for their fishing. One of the secrets of the White-breasted Kingfisher's ubiquity is that it can live on insects, lizards and other small terrestrial animals. Other accomplished fish-catchers like cormorants have 'tooth-edged' bills with which they can grip fish. Dabbling ducks have widened bills with laminations on the edges of the upper and lower mandibles. This adaptation is especially necessary for plankton-sieving shovellers, and the larger spoonbills and flamingos. Shore birds have thin, elongated bills for probing the mud in search of small animals.

Birds of prey have developed deeper, shorter and down-curved bills for tearing and piercing flesh. However, for them perhaps the most important piece of hunting equipment is their feet. They rely on the strength of their talons to kill.

Most species of bird are animal-eating, but most animals eaten are small invertebrates, in particular insects. Birds like swallows, martins and nightjars catch insects on the wing with wide gapes that scoop in their victims. Many other insects are caught on the plants on which they themselves feed. Small insectivorous birds like warblers have fine, pointed beaks for collecting them. Woodpeckers, on the other hand, have strong, dagger-shaped bills, for chiselling wood and prising insects from crevices and beneath bark.

Flowering plants also support a great number of birds, and just as birds have adapted to feeding on flowers, the trees and other plants, too, have adapted themselves to being fed on. In 1932, Salim Ali wrote on *Flower-Birds and Bird-Flowers*. He listed as characteristics of the bird-flower that pollination is possible only through birds, that they have bright and conspicuous colours (red being a bird's favourite) and no scent (as birds have a very poorly developed sense of smell), but they do have an abundant supply of nectar. Typical bird-flowers of this sort can be seen on the spectacular silk-cotton tree *Bombax malabaricum*, which is covered with waxy red or orange blooms around mid-February. Typical flower-birds are sunbirds, which have long, down-curved bills for drinking nectar.

A much larger number of bird species feeds on the fruit and seeds of trees and other plants. Finches have stout beaks built for seed crushing, while barbets and fruit-eating thrushes both have large gapes for swallowing berries whole.

HABITS – FEEDING AND BREEDING

Breeding is related very much to food supply. As large areas of the Oriental region provide more abundant food to more birds for a longer period than in more temperate zones, this means the breeding period can be longer.

Large birds of prey breed between October and March, while large waterbirds like storks, cormorants, egrets and ibis nest in colonies during the monsoon. This is also the period when munias and weaver birds breed. The peak breeding season for other common birds is February to May, although hill birds nest even later.

During the breeding season, males of many species produce long and complicated sounds called birdsong. Songs must be differentiated from the calls that birds of both sexes make throughout the year, which are much simpler and used in a variety of circumstances, for example to express alarm, to threaten or beg for food and so on. They should also be distinguished from the mechanical noises birds make, like the clattering of the bill or clapping of the wings, for similar reasons.

Once you are familiar with the song of a particular species, it is possible to identify it immediately by its voice. Not surprisingly, birds too use song to identify themselves to one another and to attract mates. In species which aggressively defend a territory, it is still debatable how much song is used simply to attract females, and to what degree it serves to demarcate the male's domain. The matter is complicated further by the fact that females are probably most attracted to the male with the largest territory. The size of territory varies enormously. In species which breed in colonies, it extends to just a metre or so, around the nest.

If song is important in finding a mate, so, too, are courtship displays. The Indian Peafowl is blessed with a powerful though unmellifluous voice, but makes up for this by the male's unmatched courtship display. The Indian Peafowl stamps and turns with his tail outspread before a number of females. Even birds which pair for life are known to display; in the breeding season, Sarus Cranes frequently break into a striking dance, spreading their wings, lowering their heads and leaping into the air, trumpeting loudly all the while. Other birds indulge in less spectacular rituals, like making gifts of food to one another.

Courtship culminates in mating, with the male mounting the female. The male's testes are internal and make up only a tiny fraction of the bird's body-weight, as little as 0.005 per cent outside the breeding season. During the breeding season the testes increase up to a thousand fold in weight. In most birds there is a small erectile phallus, but this is well developed and protrusible in only a few types like ducks and geese. The female's ovaries are also contained in her body cavity, and fertilization of an egg generally takes place once it leaves the ovary and passes down towards the shell gland where the outer layer of the egg becomes calcified to form the shell. Most females can store live sperm in their body, so fertilization can take place some time after mating.

Eggs are laid in the nest the pair has previously prepared. Nests vary from rough scrapes in the ground, as for plovers, to the holes in river banks tunnelled by kingfishers, the large untidy nests of vultures, and the elaborate nests of weaver birds often seen suspended in groups of ten or more from palms and other trees. The male weaver bird is a master of the nest-building art. Each male of a colony laboriously tears off strips of grass or leaves and proceeds to construct a retort-shaped nest, complete with entrance tunnel and egg chamber. When the nests are well under construction, the hen birds arrive and set up house in whichever nest takes their fancy. When the nest is complete and the eggs are laid, the male takes off to build another nest and attract another female.

Hornbills nest in holes in trees and are unique in imprisoning the female within the nest behind a mud wall, leaving open a gap just large enough for the male to pass her food. Not all birds form pairs. Some babblers breed communally with different females laying eggs in the same nest.

Eggs also vary in size, colour, shape and number. Those which require camouflage are well mottled, although most are pigmented. Cuckoos are parasitic and lay their eggs in others' nests, which resemble those of the host species. In the case of Hawk Cuckoos and Pied Crested Cuckoos, these hosts are babblers.

Hatchlings require different amounts of care from their parents. Songbird chicks, for example, are born naked, while game birds are born with feathers and are soon running after their parents. Most young birds have characteristic begging behaviour, and in many species parents have been observed to give them a diet richer in protein than their own normal diet, to promote their offspring's growth.

As the young gain their full plumage, the adults generally begin to moult. Feathers are shed and regrown in a distinct pattern, generally beginning with the first flight feathers and ending with the replacement of the last. Moulting is necessary as feathers do wear out. Species which have distinct breeding and non-breeding plumage moult twice a year, while in larger birds, such as eagles, a moult may take two to three years to complete. In short, individual birds shed feathers according to their needs. Migratory birds, for example, never moult at the time of migration.

MIGRATION

Many species migrate locally or over long distances to avoid adverse climatic conditions and in search of food. Hundreds of thousands of waterfowl migrate each winter to India from central and northern Asia, covering huge distances. Smaller winter migrants include wagtails, warblers and bluethroats. The huge number of wintering birds also accounts for the large numbers of wintering ornithologists and bird enthusiasts who visit the subcontinent over these months.

Summer visitors are far fewer. The multi-coloured Indian Pitta is one such bird well worth searching out. It winters in south India and Sri Lanka, and visits the deciduous forest and scrubland of the Himalayan foothills and the north-west around May, staying until it has bred.

Scientists have still to discover exactly how birds navigate during migration, although it is clear that they have a number of means to do so. Apart from sighting landmarks, it has been proved that birds make use of an internal magnetic compass and of the position of the sun and stars. It is also thought that sound and smell may play a role.

CLASSIFICATION

Like other animals, birds are classified into orders, families, genera and species. Each genus usually consists of a number of species which are obviously closely related. The first of a bird's two or three Latin names is that of its genus, and the second describes the species, while a third is used for subspecies. A species is a population of birds with a distinct identity which does not interbreed with other bird populations. Where there is a constant variation in a species, it is called a subspecies.

We begin each bird description with, first, the common English name, followed by the scientific name in italics. This is followed by the size, indicating the approximate length of the bird from the tip of its bill to the end of its tail. Then follows a description of the bird's plumage, listed separately by gender where applicable, and indications of seasonal variations of plumage, if any. After this, we have noted characteristic forms of behaviour that might help towards identification of the bird. Finally, we list the bird's food, voice, range and habitat.

BIRD TOPOGRAPHY

The scientific and English names of birds included here are based on the second edition (2011) of the *Birds of India*, by Richard Grimmett, Carol Inskipp and Tim Inskipp.

Aberrant Abnormal or unusual

Adult Mature; capable of breeding

Aerial Making use of the open sky

Aquatic Living on or in water

Arboreal Living in trees

Canopy Leafy foliage of treetops

Casque Growth above bill of hornbills

Cheek Term loosely applied to sides of head, below the eye or on ear-coverts

Collar Distinctive band of colour that encircles or partly encircles the neck

Coverts Small feathers on wings, ear and base of tail

Crepuscular Active at dusk and dawn

Crest Extended feathers on head

Crown-stripe Distinct line from forehead along centre of crown

Duars Forest areas S of E Himalayas

Ear-coverts Feathers covering the ear opening. Often distinctly coloured

Endemic Indigenous and confined to a place

Eye-ring Contrasting ring around the eye

Extinct No longer in existence

Family Specified group of genera

Forage Search for food

Flank Side of body

Foreneck The lower throat

Form Subspecies

Gape Basal part of the beak

Genus Group of related species

Jheel Shallow lake or wetland

Hackles Long and pointed neck feathers

Hepatic Rust-or liver-coloured plumage phase, mainly in female cuckoos

Iris Coloured eye membrane surrounding pupil

Lanceolate Lance-shaped; slim and pointed

Malar Stripe on side of throat

Mandible Each of the two parts of bill

Mantle Back, between wings

Mask Dark plumage round eyes and ear-coverts

Morph One of several distinct types of plumage in the same species

Moult Seasonal shedding of plumage

Nocturnal Active at night

Nominate First subspecies to be formally named

Non-passerine All orders of birds except for passerines

Nullah Ditch or stream bed, dry or wet

Order Group of related families

Paleartic Old World and arctic zone

Pelagic Ocean-going

Pied Black and white

Plumage Feathers of a bird

Primaries Outer flight feathers in wing

Race Subspecies

Range Geographical area or areas inhabited by a species

Raptors Birds of prey and vultures, excluding owls

Rump Lower back

Scapulars Feathers along edge of mantle

Secondaries Inner wing feathers

Sholas Small forests in valleys

Speculum Area of colour on secondary feathers of wings

Streamers Long extensions to feathers, usually of tail

Spangles Distinctive white or shimmering spots in plumage

Species Groups of birds reproductively isolated from other such groups

Storey Level of the tree or forest

Subspecies Distinct form that does not have specific status

Supercilium Streak above eye

Talons Strong sharp claws used to seize or kill prey

Terai Alluvial land S of Himalayas

Tarsus Lower part of a bird's legs

Terminal band Broad band on tip of feather or tail

Tertials Innermost wing-coverts, often covering secondaries

Underparts Undersurface of a bird from throat to vent

Underwing Undersurface of a wing including the linings and flight feathers

Upperparts Upper surface of a bird including wings, back and tail

Vagrant Accidental, irregular

Vent Undertail area

Wing-coverts Small feathers on wing at base of primaries and secondaries

Wingspan Length from one wing tip to the other when fully extended

Winter plumage Plumage seen during the non-breeding winter months

Black Francolin ■ *Francolinus francolinus* 35cm

DESCRIPTION Male: jet-black, spotted and marked white and fulvous; white cheeks; chestnut collar, belly and under tail-coverts. Female: browner where male is black; rufous nuchal patch; no white cheeks or chestnut collar. Solitary or small parties in high grass and edges of canals; emerges in the open in the early mornings; sometimes cocks tail. **FOOD** grain, seeds, tubers, fallen berries, insects. **VOICE** loud 3- to 6-note crow of the cock. **DISTRIBUTION** resident; N subcontinent, along foothills; south to N Gujarat and C Madhya Pradesh; an eastern race *melanotus* occurs east of Nepal in Duars; up to about 2,000m. **HABITAT** high grass, cultivation; prefers wetter areas along canals and rivers.

Grey Francolin

■ *Francolinus pondicerianus* 35cm

DESCRIPTION Sexes alike. Grey-brown and rufous above, barred and blotched; buffy-rufous below; narrow cross-bars on throat and upper breast; fine black markings on abdomen and flanks; black loop around throat encloses bright rufous-yellow throat; female smaller, with indistinct spur. Small parties, digging and moving amidst scrub and grass; seen on country roads, dust bathing or feeding; quick to take to cover on being alarmed, scattering over the area. **FOOD** cereal grains, seeds, fallen berries, insects. **VOICE** loud, high-pitched 2- to 3-note *pat...ee...la*; noisy. **DISTRIBUTION** resident: all subcontinent, south of Himalayan foothills. **HABITAT** open scrub, grass, cultivation.

Red Junglefowl
▪ *Gallus gallus* 65cm

DESCRIPTION Both sexes resemble domestic bantam breeds. Male: glistening red-orange above, with yellow about neck; metallic-black tail with long, drooping central feathers distinctive. Female: 42cm; bright chestnut forehead, supercilia continuing to foreneck; reddish-brown plumage, vermiculated with fine black and buff. Small parties, often several hens accompanying a cock; shy and skulking; emerges in clearings and on forest roads; flies up noisily when flushed. **Grey Junglefowl** G. *sonneratii* of S India has a silvery-grey plumage and gold flush on neck- and wing-coverts and is found in the low hills of the western peninsula. **FOOD** grain, crops, tubers, insects. **VOICE** characteristic crow of male, a shriller version of domestic; other cackling sounds. **DISTRIBUTION** outer Himalaya, east of Kashmir; eastern and central India, south to Narmada river. **HABITAT** *sal* forest mixed with bamboo and cultivation patches.

Grey Junglefowl

Indian Peafowl
▪ *Pavo cristatus* 110cm including full tail

DESCRIPTION Glistening blue neck and breast; wire-like crest and very long tail distinctive. Female: 85cm; lacks blue neck and breast; browner plumage; lacks the long train. Familiar bird of India; solitary or in small parties, several females with one or more males; wary in the forested parts, rather tame and confiding in many parts of W and C India around human habitation; ever-alert, gifted with keen eyesight and hearing. National bird of India. Tail feathers often illegally sold to tourists. **FOOD** seeds, berries, shoots and tubers, insects, lizards, small snakes, worms. **VOICE** loud *may-you* calls at dawn and dusk; also loud nasal calls and cackles; very noisy during the rains, when breeding. **DISTRIBUTION** all subcontinent bar Pakistan, up to about 2,000m in Himalayas. **HABITAT** forest, neighbourhood of villages and cultivated country.

Greylag Goose ■ *Anser anser* 50cm

DESCRIPTION Sexes alike. Grey-brown plumage; pink bill, legs and feet; white uppertail-coverts, lower belly and tip to dark tail; in flight, pale leading edge of wings and white uppertail-coverts distinctive. Gregarious and wary; flocks on jheels and winter cultivation; rests for most of day and feeds during night, on water and on agricultural land, especially freshly sown fields. The **Bar-headed Goose** *A. indicus* breeds in Ladakh and winters in subcontinent. **FOOD** grass, shoots of winter crops like gram and wheat; aquatic tubers. **VOICE** domestic goose-like, single-note honk, often uttered several times, loud and ringing; typical geese gaggles when feeding. **DISTRIBUTION** winter visitor; early October to mid-March; most common in N India, across the Gangetic plain to Assam, Orissa; south to N Gujarat; Madhya Pradesh; rarer south. **HABITAT** jheels, winter cultivation.

ABOVE: *Bar-headed Goose*

Ruddy Shelduck ■ *Tadorna ferruginea* 65cm

DESCRIPTION Whitish-buff head; orange-brown plumage; in flight, orangish body, white wing-coverts, green speculum and blackish flight feathers distinctive; black tail and ring

around neck (breeding). Female has a whiter head and lacks neck-ring. Young birds look like female and have some grey in wings. Pairs or small parties, rather wary; rests during day on river banks, sandbars, edges of jheels; prefers clear, open water. **FOOD** grain, shoots, insects, molluscs; reportedly also carrion. **VOICE** loud, goose-like honking, on ground and in flight. **DISTRIBUTION** breeds in Ladakh; winter visitor; all across India, less common in south. **HABITAT** rivers with sandbars, large, open jheels.

Knob-billed Duck
■ *Sarkidiornis melanotos* 75cm

DESCRIPTION Male: white head and neck, speckled black; fleshy knob (comb) on top of beak; black back has purple-green gloss; greyish lower back; white lower-neck collar and underbody; short black bars extend on sides of upper breast and flanks. Female: duller, smaller; lacks comb. Small parties, either on water or in trees over water; nests in tree cavities; feeds on surface and in cultivation; can also dive. FOOD shoots and seeds of water plants, grain, aquatic insects, worms, frogs. VOICE loud, goose-like honk when breeding. DISTRIBUTION almost entire India; mostly resident but moves considerably with onset of monsoons; uncommon in extreme S and NW India. HABITAT jheels and marshes with surrounding tree cover.

Indian Spot-billed Duck
■ *Anas poecilorhyncha* 60cm

DESCRIPTION Sexes alike. Blackish-brown plumage, feathers edged paler; almost white head and neck; black cap; dark, broad eye-stripe; green speculum bordered above with white; black bill tipped yellow; coral-red legs and feet. Pairs or small parties walk on marshy land and wet cultivation, or up-end in shallow water, usually does not associate with other ducks; when injured, can dive and remain underwater, holding on to submerged vegetation with only bill exposed. FOOD wild grain, seeds, shoots of aquatic plants; occasionally water insects, worms, water-snails; causes damage to paddy. VOICE loud, duck-like quack. DISTRIBUTION resident; N subcontinent, up to about 1,800m in Kashmir; local migrant in some areas HABITAT reed- and vegetation-covered jheels, shallow ponds.

Northern Shoveler ■ *Anas clypeata* 50cm

DESCRIPTION Broad, long beak diagnostic. Male: metallic-green head and neck; in flight, dark head, back-centre, rump and uppertail-coverts contrast with white of back and tail; also, dull-blue upperwing-coverts against dark flight feathers; metallic-green speculum and white wing bar; in overhead flight, dark head, thick white neck, dark chestnut belly and flanks. Female: mottled brown, but blue-grey shoulders (wing-coverts) and dull green speculum distinctive. Pairs or small flocks, often amidst other ducks; swims slowly, with the beak held very close to water; sometimes up-ends. **FOOD** aquatic insects, crustaceans, molluscs; vegetable matter. **VOICE** loud, 2-noted quacking notes of male. **DISTRIBUTION** winter visitor, quite common over most well-watered parts of India. **HABITAT** marshes, lakes; also vegetation-covered village ponds.

Northern Pintail ■ *Anas acuta* 60cm

DESCRIPTION A slender duck with a pointed tail. Male: greyish above; chocolate-brown head and upper neck; thin white stripe up neck side; bronze-green speculum. Female: mottled buff-brown; pointed tail lacks longer tailpins; whitish belly. Non-breeding male like female, but mantle greyish. In flight, pointed tail between feet distinctive. Highly gregarious; extremely common on vegetation-covered jheels; males often in separate flocks, especially on arrival in winter grounds; crepuscular and nocturnal; characteristic hissing swish of wings as flock flies over. **FOOD** shoots, seeds of aquatic plants, rice, also water insects and molluscs. **VOICE** usually quiet. **DISTRIBUTION** winter visitor across the subcontinent. **HABITAT** vegetation-covered jheels, lagoons.

Common Teal ▪ *Anas crecca* 32cm

DESCRIPTION Male: greyish with chestnut head with broad metallic-green band from eye to nape with yellow-white border. Black, green and buff wing speculum. Female: mottled dark and light brown; pale belly and black and green wing speculum. Most common migratory duck. Swift flier and difficult to circumvent. **FOOD** tender shoots and grains; mostly vegetarian. **VOICE** *krit...krit...*, also wheezy quack. **DISTRIBUTION** entire subcontinent in winter. **HABITAT** jheels, marshes, village ponds.

Little Grebe ▪ *Tachybaptus ruficollis* 22cm

DESCRIPTION Sexes alike. India's smallest waterbird, squat and tailless. Plumage silky and compact; dark brown above; white in-flight feathers; white abdomen. Breeding: chestnut sides of head, neck and throat; black chin; blackish-brown crown and hind-neck. Winter: white chin; brown crown and hind-neck; rufous neck. Purely aquatic; seen singly or in small, scattered groups, often diving and swimming beneath the surface. **FOOD** aquatic insects, frogs, crustaceans. **VOICE** shrill trilling notes and an occasional *click*. **DISTRIBUTION** subcontinent, up to 2,000m in Kashmir. Resident in most areas. **HABITAT** village tanks, deep jheels, lakes, reservoirs.

Juvenile

Little Egret ▪ *Egretta garzetta* 65cm

DESCRIPTION Sexes alike. A slender, snow-white waterbird. White plumage; black legs, yellow feet and black bill diagnostic. Breeding: nuchal crest of two long plumes; feathers on back and breast lengthen into ornamental filamentous feathers. The **Intermediate Egret** *Mesophoyx intermedia* is larger with black feet. The **Great Egret** *Casmerodius albus* is the largest with a noticeable kink in its neck. Small flocks feed at edge of water, sometimes wading into the shallower areas; stalks prey like typical heron, waiting patiently at edge of water. **FOOD** small fish, frogs, tadpoles, aquatic insects, crustaceans. **VOICE** an occasional croak. **DISTRIBUTION** resident; subcontinent, up to 1,600m in outer Himalaya. **HABITAT** inland marshes, jheels, riversides, damp irrigated areas; sometimes tidal creeks.

Little Egret *Intermediate Egret* *Great Egret*

Western Reef Egret ▪ *Egretta gularis* 65cm

DESCRIPTION Sexes alike. Slim with skinny neck. The pale bill is thick and heavy. Is polymorphic and varies from all white to light to dark grey. Throat usually white. Usually

solitary or seen foraging with other egrets. The rarer **Pacific Reef Egret** *E. sacra* is found only in the Andaman and Nicobar islands. **FOOD** mostly fish. **VOICE** rasping croaks. **DISTRIBUTION** W and E coasts, often inland. **HABITAT** river deltas, coasts, mangroves.

ABOVE: *Pacific Reef Egret*

Cattle Egret ■ *Bubulcus ibis* 50cm

DESCRIPTION Sexes alike. A snow-white egret seen on and around cattle and refuse heaps. Breeding: buffy-orange plumes on head, neck and back. Non-breeding: distinguished from Little Egret by yellow beak; from other egrets by size. Widespread; equally abundant around water and away from it; routinely attends to grazing cattle, feeding on insects disturbed by the animals; follows tractors; scavenges at refuse dumps and slaughter houses. Often seen on the backs of buffaloes. **FOOD** insects, frogs, lizards, refuse. **VOICE** mostly silent except for some croaking sounds when breeding. **DISTRIBUTION** resident; subcontinent, up to 1,800m in outer Himalaya. Rare in NW and NE India. **HABITAT** marshes, lakes, forest clearings, also fields.

Breeding

Purple Heron
■ *Ardea purpurea* 100cm

DESCRIPTION Sexes alike. A slender-necked, lanky bird. Slaty-purple above; black crown with long, drooping crest; rufous neck with prominent black stripe along its length; white chin and throat; deep slaty and chestnut below breast; almost black on wings and tail; crest and breast plumes less developed in female. Solitary; crepuscular; extremely shy but master of patience; freezes and hides amidst marsh reeds; when flushed, flies with neck outstretched. Active in early mornings. **FOOD** mostly fish. **VOICE** a harsh croak; sometime a high-pitched *frank*. **DISTRIBUTION** mostly resident, though numbers in some areas increase during winter because of migrants; subcontinent, south of Himalayan foothills. **HABITAT** open marshes, reed-covered lakes, riversides.

Breeding

Indian Pond Heron
■ *Ardeola grayii* 46cm

DESCRIPTION Sexes alike. A small heron, commonest of family in India; thick-set and earthy-brown in colour, with dull green legs; bill bluish at base, yellowish at centre with black tip; neck and legs shorter than in true egrets. Difficult to sight when settled; suddenly springs to notice with a flash of white wings, tail and rump. Breeding: buffy-brown head and neck; white chin and upper throat, longish crest; rich maroon back; buff-brown breast. Non-breeding: streaked dark brown head and neck; grey-brown back and shoulders; more white in plumage. Found around water, even dirty roadside puddles; ubiquitous in the plains; found in hills up to 1,200m; remains motionless in mud or up to ankles in water, or slowly stalks prey. Hunts alone; roosts in groups with other pond herons and occasionally crows. **FOOD** fish, frogs, crustaceans, insects. **VOICE** a harsh croak, usually when flushed. **DISTRIBUTION** resident; subcontinent. **HABITAT** marshes, jheels, riversides, roadside ditches, tidal creeks.

Grey Heron ■ *Ardea cinerea* 100cm

DESCRIPTION Sexes alike. A long-legged, long-necked bird of open marshes. Ashy-

grey above; white crown, neck and underparts; black stripe after eye continues as long, black crest; black-dotted band down centre of foreneck; dark blue-black flight feathers; golden-yellow iris at close range. Mostly solitary except when breeding; occasionally enters shallow water; usually stands motionless, head pulled in between shoulders, waiting for prey to come close; characteristic flight, with head pulled back and long legs trailing. **FOOD** mostly fish. **VOICE** loud *quaak* in flight. **DISTRIBUTION** mostly resident; subcontinent; breeds up to 1,750m in Kashmir. **HABITAT** marshes, tidal creeks, freshwater bodies.

Painted Stork ■ *Mycteria leucocephala* 95cm

DESCRIPTION Sexes alike. White plumage; blackish-green and white wings; blackish-green breast-band and black tail; rich rosy-pink wash on greater wing-coverts; large, slightly curved orangish-yellow bill. Young: pale dirty brown, the neck feathers edged darker; lacks breast-band. Common and gregarious; feeds with beak partly submerged, ready to grab prey; when not feeding, settles hunched up outside water; regularly soars high on thermals. **FOOD** fish, frogs, crustaceans. **VOICE** characteristic mandible clattering of storks; young in nest have grating begging calls. **DISTRIBUTION** resident and local migrant, from terai south through the area's well-watered regions. **HABITAT** inland marshes, jheels; occasionally riversides.

Woolly-necked Stork ■ *Ciconia episcopus* 105cm

DESCRIPTION Sexes alike. A large black and white stork with red legs; glossy black crown, back and breast and huge wings, the black parts having a distinct purplish-green sheen; white neck, lower abdomen and under tail-coverts; long, stout bill black, occasionally tinged crimson. In young birds, the glossy black is replaced by dark brown. Solitary or in small scattered parties, feeding along with other storks, ibises and egrets; stalks on dry land too; settles on trees. **FOOD** lizards, frogs, crabs. **VOICE** only clattering of mandibles. **DISTRIBUTION** resident; subcontinent, up to about 1,400m in the Himalaya. **HABITAT** marshes, cultivation, wet grasslands.

Black-headed Ibis ■ *Threskiornis melanocephalus* 75cm

DESCRIPTION Sexes alike. White plumage; naked black head; long, curved black bill; blood-red patches seen on underwing and flanks in flight. Breeding: long plumes over

neck; some slaty-grey in wings. Young: head and neck feathered; only face and patch around eye naked. Gregarious; feeds with storks, spoonbills, egrets and other ibises; moves actively in water, the long, curved bill held partly open and head partly submerged as the bird probes the nutrient-rich mud. **FOOD** frogs, insects, fish, molluscs, algal matter. **VOICE** loud, booming call. **DISTRIBUTION** resident; local migrant; subcontinent, from terai south. **HABITAT** marshes; riversides.

Red-naped Ibis
■ *Pseudibis papillosa* 70cm

DESCRIPTION Sexes alike. Glossy black plumage; slender, blackish-green, down-curved beak; red warts on naked black head; white shoulder-patch; brick-red legs. The **Glossy Ibis** *Plegadis falcinellus* is deep maroon-brown above, with purple-green gloss from head to lower back; feathered head and lack of white shoulder-patch distinctive. Small parties; spends most time on the drier edges of marshes and jheels; when feeding in shallow water, often feeds along with other ibises, storks, spoonbills. **FOOD** small fish, frogs, earthworms, insects, lizards, crustaceans. **VOICE** loud 2- to 3-note nasal screams, uttered in flight. **DISTRIBUTION** resident; NW India, east through Gangetic plains; south to Karnataka. **HABITAT** cultivated areas; edges of marshes.

Glossy Ibis

Spot-billed Pelican ■ *Pelecanus philippensis* 150cm

DESCRIPTION Sexes alike. Whitish plumage sullied with grey-brown; pink on lower back, rump and flanks; white-tipped brown crest on back of head; black primaries and dark brown secondaries distinctive in flight; flesh-coloured gular pouch has a bluish-purple wash; at close range the blue spots on upper mandible and on gular pouch confirm identity of species. Purely aquatic; seen singly as well as in large gatherings. Similar **Dalmation Pelican** *P. crispus* and **Great White Pelican** *P. onocrotalus* are rarer. **FOOD** mostly fish. **VOICE** mostly silent. **DISTRIBUTION** breeds in the well-watered parts of S, SE and E India but population spreads in non-breeding season. **HABITAT** large jheels, lakes.

Spot-billed Pelican

Dalmatian Pelican

Great White Pelican

Darter ■ *Anhinga melanogaster* 90cm

DESCRIPTION Sexes alike. Long, snake-like neck, pointed bill and stiff, fan-shaped tail confirm identity. Adult: black above, streaked and mottled with silvery-grey on back and wings; chocolate-brown head and neck; white stripe down sides of upper neck; white chin and upper throat; entirely black below. Young: brown with rufous and silvery streaks on mantle. A bird of deep, fresh water; small numbers scattered along with Little Cormorants; highly specialized feeder, the entire structure of the bird is modified for following and capturing fish underwater; swims low in water with only head and neck uncovered; chases prey below water with wings half open, spearing a fish with sudden rapier-like thrusts made possible by bend in neck at 8th and 9th vertebrae, which acts as a spring as it straightens. Tosses fish into air and swallows it head-first. Basks on tree stumps and rocks, cormorant style. **FOOD** mostly fish. **VOICE** loud croaks and squeaks. **DISTRIBUTION** subcontinent, south of the Himalayan foothills. **HABITAT** freshwater lakes, jheels.

Great Cormorant ■ *Phalacrocorax carbo* 80cm

DESCRIPTION Sexes alike. Breeding adult: black plumage with metallic blue-green sheen; white facial skin and throat; bright yellow gular pouch and white thigh patches; silky white

plumes on head and neck. Non-breeding adult: no white thigh patches; gular pouch less bright. First year young: dull brown above, white below. Aquatic. Not a gregarious species outside breeding season; usually one or two birds feeding close by; dives underwater in search of fish. **FOOD** fish. **VOICE** usually slient. **DISTRIBUTION** resident in most areas; subcontinent, up to 3,000m in the Himalaya. **HABITAT** jheels, lakes, mountain torrents, occasionally coastal lagoons.

Little Cormorant ■ *Phalacrocorax niger* 50cm

DESCRIPTION Sexes alike. India's smallest and most common cormorant; short, thick neck and head distinctive; lacks gular patch. The **Indian Cormorant** *P. fuscicollis* is

larger with a more oval-shaped head. Breeding adult: black plumage has blue-green sheen; silky white feathers on fore-crown and sides of head; silvery-grey wash on upper back and wing-coverts, speckled with black. Non-breeding adult: white chin and upper throat. Gregarious; flocks in large jheels; swims with only head and short neck exposed; dives often; the hunt can become a noisy,

jostling scene; frequently perches on poles, trees and rocks, basks with wings spread open. **FOOD** mostly fish; also tadpoles, crustaceans. **VOICE** mostly slient. **DISTRIBUTION** subcontinent, south of the Himalaya. **HABITAT** village tanks, jheels, lakes, occasionally

ABOVE: *Indian Cormorant*   rivers and coastal areas.

Oriental Honey-buzzard
■ *Pernis ptilorhyncus* 67cm

DESCRIPTION Sexes alike. Slender head and longish neck distinctive; tail rarely fanned. Highly variable phases. Mostly darkish brown above; crest rarely visible; pale brown underbody, with narrow whitish bars; pale underside of wings barred; broad dark subterminal tail-band; two or three more bands on tail; tarsus unfeathered. Solitary or in pairs, perched on forest trees or flying; often enters villages and outskirts of small towns. **FOOD** bee larvae, honey, small birds, lizards; occasionally robs poultry. **VOICE** high-pitched, long-drawn *weeeeeu…* **DISTRIBUTION** resident and local migrant; subcontinent to about 2,000m in the Himalaya. **HABITAT** forest, open country, cultivation, vicinity of villages.

Black-winged Kite
■ *Elanus caeruleus* 32cm

DESCRIPTION Sexes alike. Pale grey-white plumage, whiter on head, neck and underbody; short black stripe through eye; black shoulder-patches and wing-tips distinctive at rest and in flight; blood-red eyes. Young: upper body tinged brown, with pale edges to feathers. Usually solitary or in pairs, rests on exposed perch or flies over open scrub and grass country; mostly hunts on wing, regularly hovering like a kestrel to scan ground; drops height to check when hovering, with legs held ready. **FOOD** insects, lizards, rodents, snakes. **VOICE** high-pitched squeal. **DISTRIBUTION** subcontinent, up to about 1,500m in outer Himalaya. **HABITAT** open scrub and grass country; light forest.

Black Kite ■ *Milvus migrans* 60cm

DESCRIPTION Sexes alike. Dark brown plumage; forked tail, easily seen in flight; underparts faintly streaked. The **Black-eared Kite** M. *m. lineatus* breeds in Himalaya and winters in N and C India, is slightly larger and has a conspicuous white patch on the underwing, visible in overhead flight. Common and gregarious; most common near humans, thriving on the refuse generated, often amidst most crowded localities; roosts communally. **FOOD** omnivorous; refuse, dead rats, earthworms, insects, nestlings of smaller birds, poultry. **VOICE** loud, musical whistle. **DISTRIBUTION** resident; subcontinent up to about 2,200m in Himalaya, co-existing with Black-eared Kite in some localities. **HABITAT** mostly neighbourhood of humans.

Black-eared Kite

Brahminy Kite
■ *Haliastur indus* 80cm

DESCRIPTION Sexes alike. White head, neck, upper back and breast; rest of plumage a rich and rusty-chestnut; brownish abdomen and darker tips to flight feathers visible mostly in flight. Young: brown, like Black Kite, but with rounded tail. Solitary or small scattered parties; loves water; frequently scavenges around lakes and marshes; also around villages and towns. **FOOD** mostly stranded fish; also frogs, insects, lizards, mudskippers, small snakes, rodents. **VOICE** loud scream. **DISTRIBUTION** resident and local migrant; subcontinent, up to about 1,800m in the Himalaya. **HABITAT** margins of lakes, marshes, rivers, sea coasts.

Egyptian Vulture ■ *Neophron percnopterus* 65cm

DESCRIPTION Sexes alike. White plumage; blackish in wings; naked yellow head, neck and throat; yellow bill; thick ruff of feathers around neck; wedge-shaped tail and blackish flight feathers distinctive in overhead flight. The nominate race of NW India is slightly larger and has a dark horny bill. Several usually together, perched atop ruins, earthen mounds or just walking on ground; glides a lot but rarely soars high; sometimes with other vultures. **FOOD** refuse, carrion, insects, stranded turtles, the bird being specially adept at opening live turtles. **VOICE** usually silent. **DISTRIBUTION** all India; plains to about 2,000m in the Himalaya. **HABITAT** open country; vicinity of human habitation.

Juvenile

White-rumped Vulture
■ *Gyps bengalensis* 90cm

DESCRIPTION Sexes alike. Blackish-brown plumage; almost naked head has whitish ruff around base; white rump (lower back) distinctive, when perched and often in flight; in overhead flight, white underwing-coverts contrast with dark underbody and flight feathers. Young birds are brown and show no white on underwing in flight. Increasingly becoming uncommon, now rarely seen at carcasses, slaughter houses, refuse dumps. When resting, the head and neck are dug into the shoulders; soars high on thermals; several converge onto a carcass; basks in sun. **FOOD** mostly scavenges on carcasses **VOICE** loud screeches when feeding. **DISTRIBUTION** resident; all India, to about 2,800m in the Himalaya. **HABITAT** open country.

Red-headed Vulture ▪ *Sarcogyps calvus* 85cm

DESCRIPTION Sexes alike. Black plumage with white on thighs and breast; naked red head, neck and feet; in overhead flight, the white breast, thigh patches and grey-white band along wings are

distinctive; widely spread primaries. Young birds are darkish-brown with white abdomen and under tail. Mostly solitary but 2 to 4 may be seen at a carcass along with other vultures; usually does not mix with the rest. **FOOD** chiefly scavenges on carcasses. **VOICE** a hoarse croak. **DISTRIBUTION** resident; all India, up to about 2,800m in the Himalaya; uncommon. **HABITAT** open country, village outskirts.

Crested Serpent Eagle
▪ *Spilornis cheela* 75cm

DESCRIPTION Sexes alike; female larger. Dark brown plumage; roundish, pied crest, visible when erected; pale brown below, finely spotted white; in overhead flight, the dark body, white bars along the wings and white tail-band confirm identity; characteristic call. Solitary or in pairs, flying over forest, often very high, calling frequently; perches on leafy branches; swoops down on prey, snatching it in its talons; raises crest when alarmed. **FOOD** snakes, lizards, birds, rodents, squirrels. **VOICE** loud whistling scream, *keee...kee...ke...* **DISTRIBUTION** resident; subcontinent, to about 3,000m in the Himalaya. **HABITAT** forested country.

Eurasian Marsh Harrier ■ *Circus aeruginosus* 55cm

DESCRIPTION Male: dark brown plumage; dull rufous head and breast; silvery-grey wings, tail; black wing-tips (best seen in flight). Female (and young): chocolate-brown; buff on head and shoulders; like Black Kite, but tail rounded (not forked). Solitary or in pairs; sails low over a marsh, grassland or cultivation; often drops onto ground, frequently vanishing in dense grass and reed growth; perches on mounds or edges of marshes. **FOOD** fish, rodents, frogs, small waterbirds, insects. **VOICE** usually silent. **DISTRIBUTION** winter visitor; common; subcontinent, south of foothills country; most common in N India. **HABITAT** marshes, jheels, wet cultivation.

Shikra ■ *Accipiter badius* 32cm

DESCRIPTION Ashy-grey above; whitish below, close-barred with rust-brown; grey throat-stripe; in flight, the multi-banded tail and roundish wings help identification; golden-yellow eyes and yellow legs and feet seen at close range. The migrant **Eurasian Sparrowhawk** *A. nisus* is very similar but a closer look reveals the longer legs, rufous cheek-patch and absence of mesial stripe in *nisus*. Usually solitary; hides in leafy branches; pounces on unsuspecting prey; occasionally chases small birds; soars over forest. **FOOD** rodents, small birds, lizards, large insects; also robs poultry. **VOICE** loud, drongo-like *titew...titew*. **DISTRIBUTION** resident; subcontinent, up to 1,600m in the Himalaya. **HABITAT** light forest, open country, neighbourhood of villages, also in cities.

Eurasian Sparrowhawk

White-eyed Buzzard ■ *Butastur teesa* 45cm

DESCRIPTION Sexes alike. Ashy-brown above; distinct throat, white with two dark cheek-stripes and a third stripe from chin; white nape-patch, white eyes and orange-

yellow cere visible from close quarters; in flight, a pale shoulder-patch from above; from below, the pale underside of roundish wings against a darkish body distinctive. Solitary or scattered pairs; seen on exposed perches, trees, poles or telegraph wires; seems to prefer certain sites; soars high and does aerial displays when breeding. **FOOD** rodents, lizards, squirrels, small birds, frogs, insects. **VOICE** musical, plaintive *te…twee.* frequently when breeding. **DISTRIBUTION** resident; subcontinent, up to about 1,200m in Himalaya. **HABITAT** open and dry forest; cultivated country.

Greater Spotted Eagle ■ *Aquila clanga* 65cm

DESCRIPTION Sexes alike, but female slightly larger. Deep brown above, with purplish wash on back; somewhat paler below; often has whitish rump; soars on straight wings, with drooping tips; immature birds may have white markings above. The **Indian Spotted Eagle** *A. hastata* is slightly smaller, with narrower wings, and is paler above. Mostly solitary; prefers vicinity of water; perches for long spells on bare trees or on ground; sluggish behaviour. **FOOD** small animals, waterfowl, small birds. **VOICE** loud, shrill *kaek…*

kaek…, often from perch. **DISTRIBUTION** breeds sporadically in parts of N, E and NC India; spreads south in winter. **HABITAT** tree-covered areas in the vicinity of water.

BELOW: *Indian Spotted Eagle*

Crested Hawk Eagle ■ *Nisaetus cirrhatus* 70cm

DESCRIPTION Sexes alike, but female larger. Large, slender, crested forest eagle. Brown above; white underbody longitudinally streaked all over with brown; prominent occipital crest; the streaked whitish body, broad wings and long, rounded tail distinctive in flight. Solitary; occasionally a pair circles high over forests, especially when breeding; surveys for prey from high, leafy branches near forest clearings. The **Changeable Hawk Eagle** *N. c. limnaeetus*, of Himalayan foothills and NE India, is very similar except for smaller, often indistinct crest. **FOOD** partridges, other ground-birds, squirrels, hares, lizards. **VOICE** loud, screaming cry, usually long-drawn. **DISTRIBUTION** resident; subcontinent, south of the Himalaya. **HABITAT** semi-evergreen and deciduous forest, clearings.

Changeable Hawk Eagle

Common Kestrel ■ *Falco tinnunculus* 35cm

DESCRIPTION Male: black-streaked, ash-grey crown, sides of neck and nape; rufous mantle, black-spotted; cheek-stripe; grey tail has white tip and black subterminal band; streaked and spotted buffy underbody. Female: pale rufous above; streaked head and narrowly barred back; paler buff below, densely streaked. Young: like female; thickly streaked below. Solitary or in pairs; on exposed perches overlooking open country; circles in air and pounces into grass and scrub; often hovers when hunting. **FOOD** insects, lizards, small rodents. **VOICE** an infrequent clicking sound. **DISTRIBUTION** resident; local migrant; several races; breeds in the Himalaya (most common in the west); also in W Ghats south of Mumbai; associated hill ranges in S India; winter numbers augmented. **HABITAT** open country; cliffsides.

Oriental Hobby ■ *Falco severus* 28cm

DESCRIPTION Sexes alike. Small, robust falcon; slaty-grey above; deep black head, including cheeks; chestnut underparts, paler on throat. The **Eurasian Hobby** *F. subbuteo* has rusty-white underparts, thickly streaked. Solitary or several together; feeds mostly around dusk and dawn, in twilight; flies about erratically, circling, dancing, rising and dropping; charges after prey at tremendous speed; eats on wing or on perch. **FOOD** large flying insects, small bats, birds, lizards. **VOICE** shrill trill of 3- to 4-notes. **DISTRIBUTION** resident in Himalayas and NE India. **HABITAT** forested, hilly country.

ABOVE: *Eurasian Hobby*

Great Indian Bustard
■ *Ardeotis nigriceps* 120cm

DESCRIPTION Male: black crown, short crest; sandy-buff upper body, finely marked black; white below; black band on lower breast. Female: 92cm; smaller size; breast gorget broken and only rarely full. Scattered pairs or small parties; shy, difficult to approach; enters immediate vicinity of *Bishnoi* villages and other rural habitation; fast runner; hides in shade of bushes; flies low over ground. **FOOD** grain, seeds, tubers; also insects, rodents, snakes, lizards. **VOICE** loud *whonk...*, often audible for over a mile. **DISTRIBUTION** resident and local migrant; distant areas of Rajasthan, Gujarat, Maharashtra, Karnataka; numbers and erstwhile range much reduced today. **HABITAT** open grassland and scrub; semi-desert.

White-breasted Waterhen ■ *Amaurornis phoenicurus* 32cm

DESCRIPTION Sexes alike. White forehead and sides of head; dark slaty-grey above; silky white below; slaty-grey sides of breast and flanks; rufous on vent and under tail-coverts.

Solitary or in small parties; often around village ponds and tanks, occasionally derelict patches in towns; jerks stumpy tail as it walks with long strides; climbs trees easily, especially when breeding. **FOOD** insects, worms, molluscs, shoots of marsh plants. **VOICE** very noisy when breeding during rains, a series of loud croaks and chuckles, the commonest being a harsh *krr...khwakk...*; often calls through the night; silent during dry season. **DISTRIBUTION** south from Himalayan foothills. **HABITAT** reed-covered marshes, ponds, tanks, monsoon cultivation, streams.

Brown Crake
■ *Amaurornis akool* 28cm

DESCRIPTION Sexes alike, but female slightly smaller. Darkish olive-brown upper body, wings and tail; white chin and throat fade into ashy-grey underparts; browner on breast, flanks and abdomen. Solitary or pairs; mostly crepuscular; extremely elusive and secretive; feeds in early mornings and late evenings on edges of jheels, flicking its stub tail and generally moving very suspiciously. **FOOD** insects, molluscs, seeds of marshland plants. **VOICE** mostly silent, but a plaintive note and a long-drawn, vibrating whistle have been described. **DISTRIBUTION** resident and local migrant; south from Kashmir lowlands down through the peninsula, at least to Karnataka and Orissa. **HABITAT** reed-covered marshes, irrigation channels, dense growth on jheels.

Purple Swamphen ■ *Porphyrio porphyrio* 45cm

DESCRIPTION Sexes alike. Purplish-blue plumage; long red legs with oversized toes distinctive; thickish red beak; bald red forehead (casque); white under stumpy tail, seen

when tail flicked up; bald red patch on forehead smaller in female than male. Small parties amidst reeds and other vegetation on marsh and jheels; sometimes large gatherings on vegetation-covered waterbodies; walks on floating growth, rarely swims; rather tame in some areas. **FOOD** vegetable matter, seeds, tubers; known to damage paddy crops; insects, molluscs and small frogs. **VOICE** noisy when breeding, a mix of cackling and hooting notes. **DISTRIBUTION** mostly resident throughout the subcontinent, up to about 1,500m in Kashmir. **HABITAT** vegetation and reed-covered jheels, tanks.

Common Moorhen ■ *Gallinula chloropus* 32cm

DESCRIPTION Sexes alike. Dark grey head and neck; dark brownish-olive above; slaty-grey below, white centre of abdomen; fine white border to edge of wing; bright red frontal shield and base of beak with greenish-yellow tip diagnostic; greenish legs. Usually in small parties; most common in winter; moves amidst marsh vegetation, jerking tail; good swimmer; jerks head as it swims. The larger **Eurasian Coot** *Fulica atra* is a widespread resident. **FOOD** seeds and tubers of water plants, insects, molluscs, small fish and frogs. **VOICE** occasional loud, harsh *prruck...*; noisy when breeding, uttering loud croaking notes. **DISTRIBUTION** throughout the subcontinent, up to about 2,400m in Himalaya; breeds commonly in Kashmir, but also in parts of the peninsula. **HABITAT** vegetation and reed-covered ponds, tanks, jheels.

ABOVE: *Eurasian Coot*

Sarus Crane
■ *Grus antigone* 165cm

DESCRIPTION Sexes alike, but female slightly smaller than male; grey plumage; naked red head and upper neck; young birds are brownish-grey, with rusty-brown on head. Pairs, family parties or flocks; also feeds along with other waterbirds; known to pair for life and usually well-protected in northern and west-central India, but habitat loss continues to be a grave threat; flies under 12 metres off ground. **FOOD** fish, frogs, crustaceans, insects, grains, tubers. **VOICE** very loud, far-reaching trumpeting, often a duet between a pair; elaborate dancing rituals. **DISTRIBUTION** most common in N and C India (E Rajasthan, Gujarat, N and C Madhya Pradesh, Gangetic plain). **HABITAT** marshes, jheels, well-watered cultivation, village ponds.

Demoiselle Crane
■ *Grus virgo* 75cm

DESCRIPTION Sexes alike. Overall plumage grey; black head and neck; prominent white ear tufts; long black feathers of lower neck fall over breast; brownish-grey secondaries sickle-shaped and drooping over tail. Young birds have grey head and much shorter drooping secondaries over tail than adults. Huge flocks in winter, often many thousands; feeds early mornings and early evenings in cultivation; rests during hot hours on marsh edges and sandbanks; flies en masse when disturbed. **FOOD** wheat, paddy, gram; does extensive damage to winter crops. **VOICE** high-pitched, sonorous *kraak…kraak…* calls. **DISTRIBUTION** winter visitor; most common in NW India and over E Rajasthan, Gujarat and Madhya Pradesh, though sporadically over much of the area. **HABITAT** winter crop fields, sandy riverbanks, ponds, jheel edges.

Indian Thick-knee ■ *Burhinus o. indicus* 40cm

DESCRIPTION Sexes alike. Sandy-brown plumage, streaked dark; whitish below breast; thickish head, long, bare, yellow legs and large eye-goggles diagnostic; white wing-patch in flight. Solitary or in pairs; strictly a ground bird; crepuscular and nocturnal; rather quiet, sitting for long hours in same patch, where seen regularly; colouration and habitat makes it difficult to spot; squats tight or runs in short steps when located and disturbed, moving suspiciously. The **Great Thick-knee** *Esacus recurvirostris* is larger and has an upturned bill. **FOOD** small reptiles, insects, slugs; also seeds. **VOICE** a plaintive, curlew-like call at dusk and thereafter; also sharp *pick...pick...* notes. **DISTRIBUTION** drier parts of the subcontinent, up to about 1,200m in outer Himalaya. **HABITAT** light, dry forest, scrub, dry river banks, ravinous country, orchards, open *acacia* clad areas.

Great Thick-knee

Pheasant-tailed Jacana ■ *Hydrophasianus chirurgus* 30cm

DESCRIPTION Sexes alike. Male: breeding plumage chocolate-brown and white; golden-yellow on hind-neck. Dull brown and white; when not breeding, also has blackish necklace and lacks long tail; very long toes diagnostic. Solitary or in pairs when breeding; small flocks in winter; purely aquatic, moving on vegetation-covered pond surfaces; unusually

long toes enable it to walk on the lightest of floating leaves; quite confiding on village ponds. **FOOD** mostly seeds, tubers, roots; also insects, molluscs. **VOICE** loud, mewing call when breeding, two birds often calling in duet. **DISTRIBUTION** throughout area, up to about 1,500m in Kashmir; occasionally seen much higher. **HABITAT** ponds and jheels covered with floating vegetation.

River Lapwing
■ *Vanellus duvaucelii* 30cm

DESCRIPTION Sexes alike. Black forehead, crown, crest drooping over back; sandy grey-brown above; black and white wings; black chin and throat, bordered white; grey-brown breast-band; white below with black patch on belly; black spur at bend of wing. Usually pairs in close vicinity; may collect into small parties during winter, sometimes with other waders; makes short dashes or feeds at water's edge; often remains in hunched posture, when not easy to spot; slow flight; often swims and dives. **FOOD** crustaceans, insects, small frogs. **VOICE** rather like that of Red-wattled Lapwing, only a bit softer and less shrill; also a sharp *deed...did... did...* **DISTRIBUTION** breeds in parts of E and C India, including Orissa, Andhra Pradesh and eastern Madhya Pradesh; may disperse in winter. **HABITAT** stony river beds, sandbanks; sometimes collects around jheels in winter.

Yellow-wattled Lapwing ■ *Vanellus malabaricus* 28cm

DESCRIPTION Sexes alike. Jet-black cap, bordered with white; sandy-brown upper body; black band in white tail; in flight, white bar in black wings; black chin and throat; sandy-brown breast; black band on lower breast; white below; yellow lappets above and in front of eyes and yellow legs diagnostic. Solitary or in pairs, rarely small gatherings; sometimes with the more common **Red-wattled Lapwing** *V. indicus*; as a rule, prefers drier habitat, quiet and unobtrusive, feeds on ground, moving suspiciously. **FOOD** mostly insects. **VOICE** short, plaintive notes; on the whole a quiet bird; quick-repeated notes when nest site intruded upon. **DISTRIBUTION** from NW India south through the area; does not occur in extreme NE. **HABITAT** dry, open country.

Red-wattled Lapwing

Little Ringed Plover ■ *Charadrius dubius* 16cm

DESCRIPTION Sexes alike. Sandy-brown above; white forehead; black bands on head and breast and white neck-ring diagnostic; white chin and throat; lack of wing bar in flight and yellow legs and ring around eye additional clues. Small numbers, often along with

other shorebirds; runs on ground, on mud and drying jheels, walks with characteristic bobbing gait, picking food from ground; when approached close, flies rapidly, low over ground, zigzag flight accompanied by a whistling note. **FOOD** insects, worms, tiny crabs. **VOICE** a *few...few...* whistle, high-pitched but somewhat plaintive, uttered mostly on wing. **DISTRIBUTION** resident and local migrant; throughout area south from Himalayan foothills. **HABITAT** shingle-covered riverbanks, tidal mudflats, estuaries, lake edges.

Pied Avocet ■ *Recurvirostra avosetta* 45cm

DESCRIPTION Sexes alike. Black and white plumage, long, bluish legs and long, slender upcurved beak diagnostic. In flight, the long legs extend much beyond the tail. Usually

gregarious, only sometimes 2 to 3 birds scattered over waterbody; frequently enters shallow water; characteristic sideways movement of head when feeding, the head bent low as upcurved beak sweeps along bottom mud; also swims and up-ends, duck-like. **FOOD** aquatic insects, minute molluscs, crustaceans. **VOICE** loud, somewhat fluty *klooeet* or *kloeep* call, mostly on wing; also some harsh, screaming notes. **DISTRIBUTION** breeds in Kutch, N Balochistan; winter visitor, sporadically over most parts of area, most common in NW regions. **HABITAT** freshwater marshes, coastal tidal areas, creeks.

Greater Painted-snipe ■ *Rostratula benghalensis* 25cm

DESCRIPTION Polyandrous. Breeing female: metallic-olive above, thickly marked buff and black; buff stripe down crown centre; chestnut throat, breast and sides of neck; white below breast. Breeding male: duller overall; lacks chestnut. Sexes difficult to distinguish when not in breeding plumage. Crepuscular and nocturnal; solitary or a few scattered birds; feeds in squelchy mud but also moves on drier ground; runs on landing. **FOOD** insects, crustaceans, molluscs, vegetable matter. **VOICE** common call a long-drawn, mellow note that can be likened to the noise made by blowing into a bottle mouth. **DISTRIBUTION** resident throughout the area up to about 2,000m in the Himalaya. **HABITAT** wet mud, marshes, such areas where there is a mix of open water, and heavy low cover.

Common Snipe ■ *Gallinago gallinago* 28cm

DESCRIPTION Sexes alike. Cryptic-coloured marsh bird, brownish-buff, heavily streaked and marked buff, rufous and black; dull white below. Fast, erratic flight; 14 or 16 tail feathers, whitish wing-lining distinctive, but not easily seen. The **Pintail Snipe** G. *stenura* is very similar and usually distinguished only when held in the hand and with considerable experience in observation. Usually several in dense marsh growth; very difficult to see unless flushed; probes with long beak in mud, often in shallow water; feeds mostly during mornings and evenings, often continuing through the night. **FOOD** small molluscs, worms, insects. **VOICE** loud pencil call when flushed. **DISTRIBUTION** breeds in parts of W Himalaya; mostly winter visitor over the subcontinent, most common in N and C India. **HABITAT** marshlands, paddy cultivation, jheel edges.

Pintail Snipe

Black-tailed Godwit
▪ *Limosa limosa* 40cm

DESCRIPTION Sexes alike. Female slightly larger than male. Grey-brown above; whitish below; very long, straight beak; in flight, broad, white wing-bars, white rump and black tail-tip distinctive. In summer, dull rufous-red on head, neck and breast, with close-barred lower breast and flanks. The **Bar-tailed Godwit** *L. lapponica* has a slightly upcurved beak; in flight, lack of white wing-bars and barred black and white tail help identification. Gregarious, often with other large waders; quite active, probing with long beak; wades in water, the long legs often barely visible; fast and graceful, low flight. **FOOD** crustaceans, worms, molluscs, aquatic insects. **VOICE** an occasional, fairly loud *kwika...kwik*. **DISTRIBUTION** winter visitor; fairly common over N India; lesser numbers towards E and S India. The Bar-tailed is most common along the W seaboard, south to between Goa and Mumbai. **HABITAT** marshes, estuaries, creeks.

Bar-tailed Godwit

Eurasian Curlew ▪ *Numenius arquata* 58cm

DESCRIPTION Sexes alike. A large wader. Sandy-brown upper body, scalloped fulvous and black; white rump and lower back; whitish below, streaked black; very long, down-curved beak. The very similar **Whimbrel** *N. phaeopus* is smaller; has a blackish crown with white stripe through centre, and white stripes on sides of head. Mostly solitary; feeds with other large waders; runs on ground, between tidemarks, occasionally venturing into very shallow water; a truly wild and wary bird, not easy to approach close. **FOOD** crustaceans, insects, mudskippers. **VOICE** famed scream; a wild, rather musical *cour...leeor cooodee...* the first note longer. **DISTRIBUTION** winter visitor; sea coast, west to east; large inland marshes, rivers. **HABITAT** estuaries, creeks, large remote marshes.

ABOVE: *Whimbrel*

Common Redshank
■ *Tringa totanus* 28cm

DESCRIPTION Sexes alike. Grey-brown
above; whitish below, faintly marked about
breast; white rump, broad band along trailing
edge of wings; orange-red legs and base of beak.
In summer, browner above, marked black and
fulvous, and more heavily streaked below. The
Spotted Redshank *T. erythropus* is very similar
but has red at base of only the lower mandible.
Small flocks, often with other waders; makes
short dashes, probing and jabbing deep in mud;
may also enter water, with long legs completely
submerged; a rather alert and suspicious bird.
FOOD aquatic insects, crustaceans, molluscs.
VOICE quite musical, fairly loud and shrill
tleu...ewh...ewh, mostly in flight; very similar
to Common Greenshank's call, but more shrill
and high pitched. **DISTRIBUTION** breeds
in Kashmir, Ladakh; winter visitor all over
India; fairly common. **HABITAT** marshes,
creeks, estuaries.

Spotted Redshank (in breeding plumage)

Common Greenshank
■ *Tringa nebularia* 36cm

DESCRIPTION Sexes alike. Grey-brown
above; long, slightly upcurved, blackish beak;
white forehead and underbody; in flight,
white lower back, rump and absence of white
in wings diagnostic; long, greenish legs. In
summer, darker above, with blackish centres
to feathers. The **Marsh Sandpiper**
T. stagnatilis is very similar but smaller and
has distinctly longer legs; also has distinctive
call. Either solitary or small groups of two to
six birds, often with Common Redshanks and
other waders; feeds at edge of water but may
enter water to belly level. **FOOD** crustaceans,
molluscs, aquatic insects. **VOICE** wild, ringing
tew...tew...tew... **DISTRIBUTION** winter
visitor, fairly common, most of the area.
HABITAT marshes, estuaries, creeks.

Marsh Sandpiper

Wood Sandpiper ■ *Tringa glareola* 20cm

DESCRIPTION Sexes alike. Grey-brown above, closely spotted with white; slender build; white rump and tail; white below; brown on breast; no wing bar. Summer: dark olive-brown above, spotted white. The **Green Sandpiper** *T. ochropus* is stouter, more shy, much darker and glossy brown-olive above; in flight, white rump contrasts strikingly with dark upper body; blackish below wings diagnostic. Small to medium-size flocks, often with other waders; quite active, probing deep into mud or feeding at edge. **FOOD** crustaceans, insects, molluscs. **VOICE** quite noisy; sharp, trilling notes on ground; shrill, somewhat metallic *chiff...chiff* calls when flushed; sometimes a loud, sharp *tluie...* call; *T. ochropus* has distinct, wild, ringing calls when flushed. **DISTRIBUTION** winter visitor to most of the area. **HABITAT** wet cultivation, marshes, tidal creeks, mudflats.

Green Sandpiper

Common Sandpiper ■ *Actitis hypoleucos* 20cm

DESCRIPTION Sexes alike. Olive-brown above, more ash-brown and streaked brown on head and neck sides; brown rump; white below; lightly streaked brown on breast; in flight, narrow white wing bar and brown rump; white 'hook' at shoulder; In summer, darker above and speckled. One to 3 birds, either by themselves or scattered amidst mixed wader

flocks; quite active; makes short dashes, bobbing and wagging short tail; usually flies low over water, the rapid wingbeats interspersed with short glides ('vibrating flight') helping identification of the species. **FOOD** crustaceans, insects, molluscs. **VOICE** shrill *twee...tse...tse...tse...* note, usually when flushed; longish trilling song. **DISTRIBUTION** breeds in Himalayas, Kashmir to Uttarakhand to about 3,000m; winter visitor all over the area. **HABITAT** freshwater marshes, lakes, tidal areas, creeks.

Black-winged Stilt
■ *Himantopus himantopus* 25cm

DESCRIPTION Male: jet-black mantle and pointed wings; rest of plumage glossy white. Female: dark brown where male is black; black wing underside; black spots on head; duller overall in winter. Very long, pink-red legs diagnostic; extends much beyond tail in flight. Gregarious; large numbers, often along with other waders in wetlands; long legs enable it to enter relatively deep water; clumsy walk; submerges head when feeding; characteristic flight silhouette. FOOD aquatic insects, molluscs, vegetable matter. VOICE shrill notes in flight, very tern-like; noisy when breeding. DISTRIBUTION resident and local migrant over most of area, south from about 1,800m in W Himalaya. HABITAT marshes, salt pans, tidal creeks, village ponds; also riversides.

Eurasian Oystercatcher
■ *Haematopus ostralegus* 42cm

DESCRIPTION Sexes alike. Pied plumage. Black head, upperparts and breast; white below; long orange beak and pinkish legs distinctive. White on throat in winter. White rump and broad wing bar conspicuous in flight. Young birds are browner than adults. Most common on sea coasts; frequently associates with other shorebirds; runs and probes mud; beak highly specialized for feeding on molluscs. FOOD molluscs, crabs, worms. VOICE piping *kleeeep*… in flight; also a shrill whistle, often double noted, uttered on ground as well as in flight. DISTRIBUTION winter visitor, specially to the coastal regions; most common on western seaboard. HABITAT rocky and sandy coastal areas.

Indian Courser ■ *Cursorius coromandelicus* 26cm

DESCRIPTION Sexes alike. Bright rufous crown; white and black stripes above and through eyes to nape; sandy-brown above; chestnut throat and breast and black belly; long, whitish legs; in flight, dark underwings. Small parties in open country; strictly a ground bird, runs in short spurts and feeds on ground, like plovers, suddenly dipping body when disturbed, flies strongly for a short distance and lands; can fly very high. **FOOD** black beetles, other insects. **VOICE** soft, hen-like clucking call in flight, when flushed. **DISTRIBUTION** most of the area south of the Himalaya, but distribution rather patchy; absent in NE. **HABITAT** open scrub, fallow land, dry cultivation.

Small Pratincole ■ *Glareola lactea* 18cm

DESCRIPTION Sexes alike. Brown forehead; sandy-grey above; during breeding has black stripe from eye to beak; white, squarish tail, tipped black; smoky-brown underbody has a rufous wash; whiter on lower breast and abdomen; long, narrow wings and short legs. The **Collared Pratincole** G. *pratincola* is larger, with a forked tail and black loop on the throat. Gregarious; large flocks over an open expanse, close to water; very swallow-like in demeanour; strong and graceful flight over water surface, catching insects on wing; flies high in late evening. **FOOD** insects taken on wing. **VOICE** soft, but harsh call notes in flight. **DISTRIBUTION** resident and local migrant; subcontinent south of outer Himalaya, from about 1,800m. **HABITAT** large and quiet riversides, sandbars, marshy expanses, coastal swamps, tidal creeks.

ABOVE: *Collared Pratincole*

Black-headed Gull ■ *Chroicocephalus ridibundus* 45cm

DESCRIPTION Sexes alike. Winter, when in India, greyish-white plumage; dark ear patches; white outer flight feathers, with black tips. Summer breeding: coffee-brown head and upper neck, sometimes acquired just before migration. The **Brown-headed Gull** C. *brunnicephalus* is larger and has white patches (mirrors) on black wing-tips. Highly gregarious; large flocks on sea coasts, scavenges in harbours; wheels over busy seaside roads or beaches; large numbers rest on rocky ground and sand; follow boats in harbours. **FOOD** offal, fish, prawns, insects, earthworms. **VOICE** noisy; querulous *kree...ah...* screams. **DISTRIBUTION** winter visitor; most common on western seaboard; also strays inland, both on passage and for short halts. **HABITAT** sea coasts, harbours, sewage outflows, refuse dumps.

Brown-headed Gull

River Tern ■ *Sterna aurantia* 42cm

DESCRIPTION Sexes alike. Very light grey above; jet-black cap and nape, when breeding; white below; narrow, pointed wings; deeply-forked tail; bright yellow, pointed beak and red legs diagnostic. In winter, black on crown and nape reduced to flecks. Solitary or small flocks, flying about erratically; keeps to riversides, calm waters, large tanks; scans over water, plunging if possible prey is sighted; rests on river banks, noisy and aggressive, especially at nesting colonies (March to mid-June). **FOOD** fish, aquatic insects; also crabs, other crustaceans and molluscs. **VOICE** an occasional harsh, screeching note. **DISTRIBUTION** most of India; most common in N and C India. **HABITAT** inland water bodies, rivers, tanks; almost completely absent on the sea coast.

Whiskered Tern ■ *Chlidonias hybrida* 26cm

DESCRIPTION Sexes alike. Black markings on crown; silvery-grey-white plumage; long, narrow wings and slightly forked, almost squarish tail; short red legs and red beak distinctive. Summer: jet-black cap and snow-white cheeks (whiskers); black belly. At rest, closed wings extend beyond tail. Large numbers fly about a marsh or tidal creek, leisurely

but methodically, beak pointed down; dive from about 5m height but turn when just about to touch the ground, picking up insects in the process; also hunts flying insects over standing crops. **FOOD** insects, crabs, small fish, tadpoles. **VOICE** sharp, wild notes. **DISTRIBUTION** breeds in Kashmir and Gangetic plain; common in winter over the area. **HABITAT** inland marshes, wet cultivation, coastal areas, tidal creeks.

Lesser Crested Tern ■ *Thalasseus bengalensis* 45cm

DESCRIPTION Sexes alike. Greyish above, with a slight lilac wash; jet-black forehead, crown and nuchal crest in summer; whitish forehead and white-streaked crown diagnostic in winter. Blackish primaries, bright orange-yellow beak and black feet diagnostic. The **Greater Crested Tern** *T. bergii* is larger and has a white forehead all year round. Small parties out at sea, sometimes coming into coastal waters; flies leisurely between 2 and 8 metres over water, hovering occasionally; dives headlong for fish. **FOOD** fish, prawns. **VOICE** high-pitched *krrreeep*… **DISTRIBUTION** in winter over the entire Indian sea coast; possibly breeds in parts of W Pakistan coast, and some of the islets off the W coast of India. **HABITAT** open sea, coastal regions.

LEFT: *Greater Crested Tern*

Indian Skimmer ▪ *Rynchops albicollis* 40cm

DESCRIPTION Sexes alike, but female slightly smaller. Slender, pointed-winged and tern-like Pied plumage, blackish-brown above, contrasting with white underbody; white forehead, neck-collar and wing bar; diagnostic yellowish-orange beak, with much longer lower mandible; red legs. Solitary or loose flocks fly over water; characteristic hunting

style is to skim over calm waters, beak wide open, the longer projecting lower mandible partly submerged at an angle, to snap up fish on striking; many rest together on sandbars. **FOOD** chiefly fish. **VOICE** a shrill scream; twittering cries at nest colony. **DISTRIBUTION** most common in N and C India, east to Assam; less common south of Maharashtra, N Andhra Pradesh. **HABITAT** large rivers; fond of placid waters.

Chestnut-bellied Sandgrouse ▪ *Pterocles exustus* 28cm

DESCRIPTION Male: sandy-buff above, speckled brown and dull yellow; black gorget and chocolate-black belly. Female: buffy above, streaked and barred darker; black-spotted breast; rufous and black-barred belly and flanks. Pointed central tail feathers and black wing-underside distinctive in flight. Huge gatherings at waterholes in dry season; regularly arrives at water; strictly a ground bird, squatting tight or shuffling slowly; rises en masse. **FOOD** seeds of grasses and weeds. **VOICE** deep, clucking *kut...ro...* call note, uttered mostly on wing. **DISTRIBUTION** all India except NE and extreme south; most common in NW and C India. **HABITAT** open areas, semi-desert, fallow land.

Yellow-footed Green Pigeon ■ *Treron phoenicopterus* 33cm

DESCRIPTION Male ashy olive-green above; olive-yellow collar, band in dark slaty tail; lilac-red shoulder-patch (mostly absent in female); yellow legs and underbody. Female

slightly duller than male. The nominate (northern) race has grey lower breast and belly. Small flocks; mostly arboreal, rarely coming to salt-licks or cropland; remains well hidden in foliage but moves briskly; has favourite feeding trees. **FOOD** fruits, berries. **VOICE** rich, mellow whistling notes. **DISTRIBUTION** south roughly of line from S Rajasthan to N Orissa to Sri Lanka; rarer in Pakistan. **HABITAT** forests, orchards, city parks, cultivated village vicinities.

Nilgiri Wood Pigeon ■ *Columba elphinstonii* 42cm

DESCRIPTION Sexes alike. Reddish-brown above; metallic purple-green on upper back; grey head and underbody; whitish throat; black and white chessboard on hind-neck diagnostic. Solitary or in small gatherings; arboreal but often descends to forest floor to pick fallen fruit; strong flier, wheeling and turning amidst branches at a fast speed; occasionally along with other frugivorous birds. **FOOD** fruits, berries, flower buds. **VOICE** loud *who* call, like a softer version of a langur's call, followed by 3 to 5 deep and eerie sounding *who…who…who…* notes; characteristic call of heavy W Ghats forest. **DISTRIBUTION** W Ghats south from Mumbai. **HABITAT** moist evergreen forest; sholas; cardamom plantations.

Eurasian Collared Dove
■ *Streptopelia decaocto* 32cm

DESCRIPTION Sexes alike. Greyish-brown plumage; lilac wash about head and neck; black half-collar on hind-neck diagnostic; broad whitish tips to brown tail feathers, seen as a terminal band when fanned during landing; dull lilac breast and ashy-grey underbody. Small parties when not breeding; often associates with other doves; large gatherings glean in cultivated country; strong flier, chases intruders in territory. **FOOD** seeds, grain. **VOICE** characteristic *kukkoo.. kook…*, almost dreamlike in quality; also a strident *koon…koon…* when male displays at onset of breeding. **DISTRIBUTION** most of the area, except extreme NE Himalaya; resident and local migrant; most common in NW, W and C India. **HABITAT** cultivation, open scrub, dry forest.

Spotted Dove
■ *Stigmatopelia chinensis* 30cm

DESCRIPTION Sexes alike. Grey and pink-brown above, spotted white; white-spotted black hind-neck collar (chessboard) diagnostic; dark tail with broad white tips to outer feathers seen in flight; vinous-brown breast, merging into white on belly. Young birds are barred above and lack chessboard. Pairs or small parties on ground; frequently settles on paths and roads, flying further on intrusion; quite tame and confiding in many areas; drinks often; at harvest times, seen along with other doves in immense gatherings. **FOOD** grains, seeds. **VOICE** familiar bird sound of India, a soft, somewhat doleful *crook…cru…croo* or *croo…croo. croo*. **DISTRIBUTION** subcontinent, up to about 3,500m in the Himalaya. **HABITAT** open forest, scrub, habitation, cultivation.

Laughing Dove
■ *Stigmatopelia senegalensis* 26cm

DESCRIPTION Sexes alike. Pinkish grey-brown plumage with black-and-white chessboard on sides of foreneck; white tips to outer-tail feathers and broad grey wing-patches best seen in flight; small size distinctive. Pairs or small flocks; associates freely with other doves in huge gatherings at harvest time; feeds mostly on ground, walking about silently. **FOOD** grains, grass, weeds and seeds. **VOICE** somewhat harsh but pleasant *cru…do…do…do…do*. **DISTRIBUTION** almost entire India up to about 1,200m in the outer Himalaya; uncommon in NE states. **HABITAT** open scrub, cultivation, neighbourhood of habitation.

Emerald Dove ■ *Chalcophaps indica* 26cm

DESCRIPTION Sexes alike. Bronze emerald-green upper body; white forehead and eyebrows; grey crown and neck; white on wing shoulder and across lower back; whitish rump diagnostic in flight; rich pinkish-brown below; coral-red beak and pink-red legs.

Solitary or in pairs; moves on forest paths and clearings or darts almost blindly through trees, usually under 5m off ground; difficult to spot on ground. **FOOD** seeds, fallen fruit; known to eat termites. **VOICE** deep, plaintive *hoo… oon…hoo…oon…*, many times at a stretch. **DISTRIBUTION** almost throughout subcontinent up to about 2,000m. Absent in Pakistan. **HABITAT** forest, bamboo, clearings; foothills.

Red Collared Dove ▪ *Streptopelia tranquebarica* 22cm

DESCRIPTION Male: deep ashy-grey head; black hind-neck collar; rich wine-red back; slaty grey-brown lower back, rump and uppertail; whitish tips to all but central tail feathers. Female: much like Eurasian Collared Dove, but smaller size and more brownish colouration distinctive. Solitary, in pairs or small parties; associates with other doves but is less common; feeds on ground, gleaning on harvested croplands; perches and suns on leafless branches and overhead wires. **FOOD** grass and other seeds, cereals. **VOICE** quick repeated *gru…gurgoo…* call, with more stress on first syllable. **DISTRIBUTION** throughout the area, south of the Himalayan foothills. **HABITAT** cultivation, scrub, deciduous country.

Rose-ringed Parakeet ▪ *Psittacula krameri* 42cm

DESCRIPTION Male: grass-green plumage; short, hooked, red beak; rosy-pink and black collar distinctive (obtained only during third year). Female: lacks the pink-and-black collar; instead, pale emerald-green around neck. Gregarious; large flocks of this species are a familiar sight in India; causes extensive damage to standing crops, orchards and garden fruit trees; also raids grain depots and markets; large roosting colonies, often along with mynas and crows. **FOOD** fruit, crops, cereal. **VOICE** shrill *keeak…* screams. **DISTRIBUTION** subcontinent, south of Himalayan foothills. **HABITAT** light forest, orchards, towns, villages.

Plum-headed Parakeet ■ *Psittacula cyanocephala* 35cm

DESCRIPTION Male: yellowish-green plumage; plum-red head; black and bluish-green collar; maroon-red wing shoulder-patch; white tips to central tail feathers distinctive. Female: dull, greyer head; yellow collar; almost non-existent maroon shoulder-patch. Pairs

or small parties; arboreal, but descends into cultivation in forest clearings and outskirts; sometimes huge gatherings in cultivation; strong, darting flight over forest. **FOOD** fruits, grain, flower nectar and petals. **VOICE** loud, interrogative *tooi...tooi...* notes in fast flight; also other chattering notes. **DISTRIBUTION** subcontinent south of Himalayan foothills. **HABITAT** forest, orchards, cultivation in forest.

Alexandrine Parakeet ■ *Psittacula eupatria* 52cm

DESCRIPTION Male: rich grass-green plumage; hooked, heavy red beak; deep red shoulder-patch; rose-pink collar and black stripe from lower mandible to collar distinctive. Female: smaller and lacks the collar and black stripe. Yellow under tail in both sexes. Both small flocks and large gatherings; feeds on fruiting trees in orchards and on standing crops, often causing extensive damage; strong flier; roosts along with other birds at favoured sites. **FOOD** fruits, vegetables, crops, seeds. **VOICE** high-pitched *kreeak...* scream, on wing as well as on perch; popular cage bird, learning to imitate some notes and few human words. **DISTRIBUTION** almost throughout the area, south of Himalayan foothills. **HABITAT** forest, orchards, cultivated areas, towns.

Sirkeer Malkoha
▪ *Taccocua leschenaultii* 45cm

DESCRIPTION Sexes alike. Olive-brown plumage; long, graduated tail, with broad white tips to blackish outer feathers diagnostic in flight; cherry-red beak, with yellow tip. Solitary or in pairs; sometimes 4 or 5 birds in the neighbourhood; move mostly on ground, in dense growth; may clamber out on some bush tops or low trees; flight weak and short. **FOOD** insects, fallen fruit, lizards. **VOICE** fairly loud and sharp clicking notes: mostly vocal when breeding. **DISTRIBUTION** most of subcontinent, up to about 1,800m in the Himalaya; absent in NW India and Kashmir. **HABITAT** open jungle, scrub, ravines, dense growth around habitation. Runs low and rat-like when disturbed. Builds its own nest. Endemic resident of thorn scrub and semi-desert regions.

Common Hawk Cuckoo
▪ *Hierococcyx varius* 35cm

DESCRIPTION Sexes alike. Ashy-grey above; dark bars on rufescent-tipped tail; dull white below, with pale ashy-rufous on breast; barred below. Young birds broadly-streaked dark below; pale rufous barrings on brown upper body. Solitary, rarely in pairs; strictly arboreal; noisy during May–September; silent after rains. **FOOD** chiefly insects; rarely wild fruit and small lizards. **VOICE** famous call notes; interpreted as *brain-fever...*, uttered untiringly in crescendo; also described as *pipeeha... pipeeha...*; very noisy in overcast weather. **DISTRIBUTION** subcontinent, south of Himalayan foothills, uncommon, even during rains, in arid zones. **HABITAT** forests, open country, near habitation.

Jacobin Cuckoo
■ *Clamator jacobinus* 33cm

DESCRIPTION Sexes alike. Black above; noticeable crest; white in wings and white tip to long tail feathers diagnostic in flight; white underbody. Young birds, seen in autumn, are dull sooty-brown with indistinct crests; white areas dull fulvous. Solitary or in small parties of 4 to 6; arboreal; occasionally descends to ground to feed on insects; arrives just before SW monsoon by end of May; noisy and active, chasing one another; mobbed by crows on arrival. **FOOD** insects, including hairy, noxious caterpillars. **VOICE** noisy; loud, metallic *plew...piu...* call notes; other shrill shrieks. **DISTRIBUTION** chiefly SW monsoon breeding visitor; most of the area south of outer Himalaya. **HABITAT** open forest, cultivation, orchards.

Asian Koel ■ *Eudynamys scolopaceus* 42cm

DESCRIPTION Male: metallic-black plumage; greenish beak and crimson eyes. Female: dark brown, thickly spotted and barred white; whitish below, dark-spotted on throat, barred below. Solitary or in pairs; arboreal; mostly silent between July and February; fast

flight. **FOOD** *ficus* and other fruits; insects, snails, eggs of smaller birds. **VOICE** familiar call of Indian countryside. Very noisy between March and June, coinciding with breeding of crows; loud *kuoo...kuooo...* whistling calls in crescendo by male koel, the first syllable longish; water-bubbling call of female. **DISTRIBUTION** subcontinent, up to about 1,800m in outer Himalaya; uncommon in drier areas. **HABITAT** light forests, orchards, city parks, cultivation, open areas.

Indian Cuckoo

■ *Cuculus micropterus* 32cm

DESCRIPTION Sexes alike. Slaty-brown above; greyer on head, throat and breast; whitish below, with broadly spaced black cross-bars; broad subterminal tail-band (characteristic of the non-hawk cuckoos of genus *Cuculus*); the female often has a rufous-brown wash on the throat and breast; call notes most important identification clue. Solitary; arboreal, not easy to see; overall appearance very hawk-like, but distinctly weaker-looking flight. The **Eurasian Cuckoo** *C. canorus* differs from the Indian Cuckoo by lacking the subterminal black band and has the diagnostic *cuck-koo* call. **FOOD** insects, with special fondness for hairy caterpillars. **VOICE** very distinct call; a 4-noted mellow whistle, variously interpreted, the best known being *bo… ko…ta…ko* or *crossword…puzzle*; the third note trailing slightly and the fourth a little more; very vocal between April and August coinciding with the breeding of its principal hosts, drongos and orioles; may call for several minutes continuously, often throughout the day if overcast. **DISTRIBUTION** subcontinent south from Himalaya to about 2,500m, excepting the drier and arid parts of NW India; absent in Pakistan. **HABITAT** forest, orchards.

Eurasian Cuckoo

Greater Coucal

■ *Centropus sinensis* 50cm

DESCRIPTION Sexes alike. Glossy bluish-black plumage, chestnut wings; blackish, loose-looking, long, graduated tail. Female somewhat bigger than male. Solitary or in pairs; moves amidst dense growth, fanning and flicking tail often; clambers up into trees, but is a poor flier, lazily flying short distances. **FOOD** insects, lizards, frogs, eggs and young of other birds, small snakes. **VOICE** loud and resonant *coop…coop…coop…* call familiar; occasionally a squeaky call. **DISTRIBUTION** subcontinent, from outer Himalaya to about 2,000m. **HABITAT** forest, scrub, cultivation, gardens, derelict patches, vicinity of habitation.

Collared Scops Owl ■ *Otus (bakkamoena) lettia* 25cm

DESCRIPTION Sexes alike. Small ear tufts and upright posture. Greyish-brown above, profusely marked whitish; buffy nuchal collar diagnostic; buffy-white underbody, streaked and mottled dark. The very similar **Indian Scops Owl** O. *bakkamoena* is mostly

distinguished by its call. Solitary or in pairs; remains motionless during day in thick, leafy branches or at junctions of stems and branches; very difficult to spot; flies around dusk. **FOOD** insects, small lizards, rodents; also small birds. **VOICE** a single-note *wut… wut…*, rather questioning in tone; calls through the night, often for 20 minutes at stretch, a *wut…*every 2 to 4 seconds. **DISTRIBUTION** resident in Himalayas and NE India. **HABITAT** forests, cultivation, orchards, trees in vicinity of habitation.

ABOVE: *Indian Scops Owl*

Indian Eagle Owl ■ *Bubo b.bengalensis* 56cm

DESCRIPTION Sexes alike. Brown plumage, mottled and streaked dark and has light; prominent ear tufts; orange eyes; legs fully feathered. The **Brown Fish Owl** *Ketupa zeylonensis* (56cm) is darker and has a white throat-patch and naked legs. Solitary or pairs; mostly nocturnal; spends day in leafy branch, rock ledge or an old well; flies slowly but considerable distances when disturbed; emerges to feed around sunset, advertising its arrival with its characteristic call. **FOOD** rodents; also reptiles, frogs and medium-sized birds. **VOICE** deep, booming *bu…boo…* call; snapping calls at nest. **DISTRIBUTION** throughout area, up to about 1,500m in the Himalaya. **HABITAT** ravines, cliffsides, riversides, scrub, open country.

ABOVE: *Brown Fish Owl*

Jungle Owlet ▪ *Glaucidium radiatum* 20cm

DESCRIPTION Sexes alike. Lacks ear tufts. Darkish brown above, barred rufous and white; flight feathers barred rufous and black; white moustachial stripe, centre of breast and abdomen; remainder of underbody barred dark rufous-brown and white. The **Asian Barred Owlet** G. *cuculoides* (23cm) of the Himalaya is slightly larger and has abdominal streaks. Solitary or in pairs; crepuscular, but sometimes also active and noisy by day; otherwise spends day in leafy branch; flies short distance when disturbed. **FOOD** insects, small birds, lizards, rodents. **VOICE** noisy; musical *kuo… kak…kuo…kak…* call notes, rising in crescendo for a few seconds only to end abruptly; other pleasant, bubbling notes. **DISTRIBUTION** throughout area, up to 2,000m in Himalaya; absent in extreme NE states. **HABITAT** forest; partial to teak and bamboo mixed forests.

Asian Barred Owlet

Spotted Owlet
▪ *Athene brama* 20cm

DESCRIPTION Sexes alike. No ear tufts. Greyish-brown plumage, spotted white. Yellowish eyes; broken whitish-buff nuchal collar. Young birds more thickly-marked white; darkish streaks below breast. Pairs or small parties; roosts during day in leafy branches, tree cavities or cavities in walls; active in some localities during daytime; disturbed birds fly to neighbouring tree or branch and bob and stare at intruder **FOOD** insects, small rodents, lizards, birds. **VOICE** assortment of scolding and cackling notes, screeches and chuckles. **DISTRIBUTION** throughout area, up to about 1,800m in outer Himalaya. **HABITAT** open forests, orchards, cultivation, vicinity of habitation.

Savanna Nightjar ■ *Caprimulgus affinis* 25cm

DESCRIPTION Male: grey-brown plumage, mottled dark; a buffy 'V' on back, from shoulders to about centre of back; two pairs of outer tail feathers white, with pale dusky tips; white wing-patches. Female: like male, but without white outer tail feathers, which are barred; conspicuous rufous-buff wing-patches; call most important identification clue. Solitary or several scattered over an open expanse; overall behaviour like that of other nightjars; remains motionless during day on open rocky, grass or scrub-covered ground;

sometimes roosts on tree, along length of a branch; flies around dusk, often flying high; drinks often. **FOOD** flying insects. **VOICE** calls on wing as well as on perch, a fairly loud, penetrating *sweeesh* or *schweee…* **DISTRIBUTION** throughout area, south of outer Himalaya to about 2,000m; moves considerably locally. **HABITAT** rocky hillsides, scrub and grass country, light forests, dry streams and river beds, fallow land, cultivation.

Jungle Nightjar ■ *Caprimulgus indicus* 32cm

DESCRIPTION Plumage in nightjars highly obliterative. Mottled and vermiculated grey-brown, black, buff and white; in some species, white tip to tail in male; calls highly diagnostic. Solitary or several scattered; crepuscular and nocturnal; squats during day, along a branch's length or on rocky ground amidst dry leaves; extremely difficult to spot unless almost stepped upon; flies around dusk, hawking insects in zigzag flight; settles on cart tracks and roads, where eyes gleam in vehicle headlights. The **Large-Tailed Nightjar** C. *macrurus* is slightly larger (33cm) and has more brownish plumage. **FOOD** winged insects. **VOICE** somewhat whistling *chuckoo…chuckoo*, up to 7 minutes at a stretch, with pauses in between;

a quick-repeated, mellow *tuck…tuck…tuck* call, 8 to 50 at a stretch; occasionally a pleasant *uk…kukrooo…* vocal between dusk and dawn. Calls help identification. **DISTRIBUTION** Resident and summer visitor from E Rajasthan to Bihar and Orissa, W peninsula. **HABITAT** forest clearings, broken scrubby ravines.

Large-tailed Nightjar

Crested Treeswift ■ *Hemiprocne coronata* 23cm

DESCRIPTION Male: bluish-grey above, with a faint greenish wash; chestnut sides of face and throat; ashy-grey breast, whiter below. Female: like male, but lacks chestnut on head. Backward-curving crest and long, deeply forked tail diagnostic. Pairs or small, scattered parties; fly during day, hawking insects; have favourite foraging areas; flight graceful, not as fast as other swifts, but displaying typical swift mastery; calls from perch and in flight; unlike other swifts, perches on bare, higher branches; drinks in flight from surfaces of forest

pools. **FOOD** winged insects. **VOICE** double-noted faint scream; also a parrot-like *kea...kea...* call. **DISTRIBUTION** subcontinent, south of Himalayan foothills; absent in the arid parts of NW India. **HABITAT** open, deciduous forest.

Little Swift ■ *Apus affinis* 15cm

DESCRIPTION Sexes alike. Blackish plumage; white rump and throat diagnostic; short, square tail and long, sickle-like swift wings. The large **Fork-tailed Swift** A. *pacificus* (18cm) has a deeply forked tail. Highly gregarious; on the wing during day, hawking insects, flying over human habitation, cliffs and ruins; strong fliers, exhibiting great mastery and control in fast wheeling flight; frequently utters squealing notes on the wing; retires to safety of nest colonies in overcast weather. **FOOD** winged insects.

VOICE musical squeals on the wing; very vocal at sunset, but also through the day. **DISTRIBUTION** throughout the area, up to about 2,400m in Himalaya. **HABITAT** human habitation, cliffs, ruins.

Fork-tailed Swift

Malabar Trogon ▪ *Harpactes fasciatus* 30cm

DESCRIPTION Male: sooty-black head and neck and breast; yellow-brown back; black wings narrowly barred white; rich crimson underbody; white breast gorget. Female: duller overall; lacks black on head and breast; orange-brown underbody. Long, squarish tail diagnostic. Solitary or in pairs; strictly arboreal; difficult to see because duller back is mostly turned towards observer or intruder; hunts flycatcher-style or flits amongst taller branches; flicks tail and bends body when disturbed. **FOOD** chiefly insects; also fruits. **VOICE** diagnostic, often a giveaway to bird's presence in a forest; 3- to 8-noted, somewhat whistling *cue… cue…* calls. **DISTRIBUTION** forested areas of peninsular India; Satpura range, W Ghats, east to Orissa and parts of E Ghats. **HABITAT** forest.

Indian Roller

▪ *Coracias benghalensis* 31cm

DESCRIPTION Sexes alike. Pale greenish-brown above; rufous-brown breast; deep blue tail has light blue subterminal band; in flight, bright Oxford-Blue wings and tail, with Cambridge-Blue bands distinctive. Solitary or in pairs; perches on overhead wires, bare branches, earthen mounds and small bush tops; either glides and drops on prey or pounces suddenly; batters prey against perch before swallowing. **FOOD** mostly insects; catches small lizards, frogs, small rodents and snakes. **VOICE** usually silent; occasionally harsh *khak…kak…kak…* notes; exuberant screeching notes and shrieks during courtship display, diving, tumbling and screaming wildly. **DISTRIBUTION** almost entire subcontinent, south of outer Himalaya, where found up to about 1,500m. **HABITAT** open country, cultivation, orchards, light forests.

Stork-billed Kingfisher
■ *Pelargopsis capensis* 38cm

DESCRIPTION Sexes alike. Solitary, more heard than seen. Does not normally hover. Enormous red bill diagnostic. Head dark grey-brown with yellowish collar on back of neck. Body pale green-blue above and brownish-yellow below. **FOOD** fish, frogs, small birds. **VOICE** Noisy *Kee…kee…kee* repeated many times. **DISTRIBUTION** subcontinent except drier parts of NW. **HABITAT** Canals, streams, coastal backwaters in well-wooded country.

White-throated Kingfisher ■ *Halcyon smyrnensis* 28cm

DESCRIPTION Sexes alike. Chestnut-brown head, neck and underbody below breast; bright turquoise-blue above, often with greenish tinge; black flight feathers and white wing-patch in flight; white chin, throat and breast distinctive; coral-red beak and legs. Solitary or scattered pairs atop overhead wires, poles and tree-tops; frequently found far from water; drops on to ground to pick up prey. **FOOD** Insects, frogs, lizards, small rodents; only occasionally fish. **VOICE** noisy; loud, crackling laugh, often audible over crowded urban areas; song a longish, quivering whistle, sounding as *kililililili…* characteristic feature of hot season, when bird is breeding; fascinating courtship display. **DISTRIBUTION** subcontinent, south of outer Himalaya. **HABITAT** forest, cultivation, lakes, riversides; also coastal mangroves and estuaries.

Common Kingfisher ■ *Alcedo atthis* 18cm

DESCRIPTION Sexes alike. Bright blue above, greenish on wings; top of head finely banded black and blue; ferruginous cheeks, ear-coverts and white patch on sides of neck; white chin and throat and deep ferruginous underbody distinctive; coral-red legs and

blackish beak. Solitary or in scattered pairs; never found away from water; perches on pole or overhanging branch; flies low over water, a brilliant blue streak, uttering its shrill notes; sometimes tame and confiding; dives for fish from perch; occasionally hovers over water before diving. **FOOD** fish; occasionally tadpoles and aquatic insects. **VOICE** shrill *chichee chichee* **DISTRIBUTION** subcontinent, south of 2,000m in Himalaya; various races differ in shade of blue-green upper body. **HABITAT** streams, lakes, canals; also coastal areas.

Pied Kingfisher ■ *Ceryle rudis* 30cm

DESCRIPTION Speckled black and white plumage diagnostic; black nuchal crest; double black gorget across breast in male. The female differs in having a single, broken breast gorget. Solitary, in pairs or in small groups; always around water, perched on poles, tree stumps or rocks; hovers when hunting, bill pointed down as wings beat rapidly; dives fast, headlong on sighting fish; batters catch on perch; calls in flight. The **Crested Kingfisher** *Megaceryle lugubris* of Himalayan streams and rivers can be identified by larger size (41cm), larger crest and white nuchal collar. **FOOD** chiefly fish; occasionally tadpoles and water insects. **VOICE** piercing, twittering *chirrruk…chirruk…* cries in flight, sounding as if the bird is complaining. **DISTRIBUTION** subcontinent, up to about 2,000m in Himalaya. **HABITAT** streams, rivers, ponds; sometimes coastal areas.

ABOVE: *Crested Kingfisher*

Green Bee-eater ▪ *Merops orientalis* 21cm

DESCRIPTION Sexes alike. Bright green plumage;
red-brown wash about head; pale blue on chin and
throat, bordered below by black gorget; slender, curved
black beak; rufous wash on black-tipped flight feathers;
elongated central tail feathers distinctive. Small parties;
perches freely on bare branches and overhead telegraph
wires; attends to grazing cattle, along with drongos, cattle
egrets and mynas; also seen in city parks and garden;
launches graceful sorties after winged insects; batters prey
against perch before swallowing. **FOOD** mostly winged
insects; confirmed nuisance to the honey industry. **VOICE**
noisy; cheerful trilling notes, chiefly uttered on wing.
DISTRIBUTION subcontinent, south of about 1,800m in
outer Himalaya. **HABITAT** open country and cultivation;
light forests.

Blue-tailed Bee-eater ▪ *Merops philippinus* 30cm

DESCRIPTION Sexes alike. Elongated central tail feathers. Greenish above, with faint
blue wash on wings; bluish rump, tail diagnostic; yellow upper throat-patch with chestnut
throat and upper breast; slightly curved black beak, broad black stripe through eye. The
very similar **Blue-cheeked Bee-eater** M. *persicus* (31cm) has a dull-white and blue-green
cheek-patch. In good light, the greenish rump and tail help identification. Usually small
flocks, frequently in vicinity of water; launches short,
elegant flights from wire or tree perch; characteristic flight,
a few quick wingbeats and a stately glide. **FOOD** winged
insects. **VOICE** musical, ringing notes, chiefly uttered in
flight. **DISTRIBUTION**
exact range of these species
not correctly known;
breeds in parts of NW
and N India, and perhaps
patchily through E and
SC India; spreads wide
during the rains and winter;
both species frequently
seen together in winter.
HABITAT open country,
light forests, vicinity of
water, cultivation; may
occasionally be seen in
coastal areas.

Blue-cheeked Bee-eater

Chestnut-headed Bee-eater
■ *Merops leschenaulti* 21cm

DESCRIPTION Sexes alike. Grass-green plumage; chestnut-cinnamon crown, hind-neck and upper back; yellow chin and throat; rufous and black gorget. Small gatherings on telegraph wires or bare upper branches of trees from where the birds launch short aerial sallies; fast, graceful flight; noisy when converging at roosting trees. **FOOD** chiefly winged insects, captured in flight. **VOICE** musical twittering notes, mostly uttered on the wing, and sometimes from perch. **DISTRIBUTION** disjunct. Himalayan foothills country, from Uttarakhand to extreme NE; a second population exists in the W Ghats south of Goa; also Sri Lanka. Occasionally may be encountered in the peninsula, especially during the monsoon. **HABITAT** vicinity of water in forested areas.

Common Hoopoe ■ *Upupa epops* 31cm

DESCRIPTION Sexes alike. Fawn-coloured plumage; black and white markings on wings, back and tail; black and white-tipped crest; longish, gently curved beak. Solitary or in scattered pairs; small, loose flocks in winter; probes ground with long beak, sometimes

feeding along with other birds; flits among tree branches; crest often fanned open; becomes rather aggressive with onset of breeding season. **FOOD** insects caught on ground or pulled from underground. **VOICE** pleasant, mellow *hoo…po… po…*, sometimes only first two notes; calls have a slightly ventriloquistic quality; calls frequently when breeding. **DISTRIBUTION** subcontinent, up to about 5,500m in Himalaya; several races; spreads considerably in winter. **HABITAT** meadows, open country, garden lawns, open light forests.

Indian Grey Hornbill ■ *Ocyceros birostris* 60cm

DESCRIPTION Grey-brown plumage; large, curved beak with casque diagnostic; long, graduated tail, tipped black and white. Casque smaller in female. The **Malabar Grey Hornbill** O. *griseus* (58cm), restricted to the W Ghats, south of Khandala, lacks casque on beak; dark tail tipped white, except on central feathers. Pairs or small parties; sometimes large gatherings; mostly arboreal, but descends to pick fallen fruit or lizards; feeds along with frugivorous birds on fruiting trees; noisy, undulating flight. **FOOD** fruit, lizards, insects, rodents. **VOICE** noisy; normal call a shrill squealing note; also other squeals and screams.

DISTRIBUTION almost throughout India, up to about 1,500m in Himalaya; absent in arid NW regions and the heavy rainfall areas of southern W Ghats. **HABITAT** forests, orchards, tree-covered avenues, vicinity of habitation.

Malabar Grey Hornbill

Great Hornbill ■ *Buceros bicornis* 130cm

DESCRIPTION Sexes alike. Black face, back and underbody; two white bars on black wings; white neck, lower abdomen and tail; broad black tail-band; huge black and yellow beak with enormous concave-topped casque distinctive. Female slightly smaller. Pairs or small parties; occasionally large flocks; mostly arboreal, feeding on fruiting trees, plucking fruit with tip of bill, tossing it up, catching it in the throat and swallowing it; may settle on ground to pick up fallen fruit; noisy flight, audible from over a kilometre away, even when flying very high, caused by drone of air rushing through bases of outer quills of wing feathers; flight: alternation of flapping and gliding, less undulating than in other hornbills. **FOOD** fruits, lizards, rodents, snakes. **VOICE** loud and deep barking calls; loud *tokk* at feeding sites, audible for considerable distance. **DISTRIBUTION** lower Himalaya east of Uttarakhand, up to about 1,800m; another population exists in W Ghats, south of Khandala. **HABITAT** forests.

Oriental Pied Hornbill ■ *Anthracoceros albirostris* 88cm

DESCRIPTION Sexes alike. Female slightly smaller. Black above; white face-patch, wing-tips (seen in flight) and tips to outer tail feathers; black throat and breast; white below. Black and yellow beak with large casque. The **Malabar Pied Hornbill** A. *coronatus* (92cm) is very similar, except for completely white outer tail feathers. Small parties, occasionally collecting into several dozen birds on favourite fruiting trees; associates with other birds;

arboreal but often feeds on ground, hopping about. **FOOD** fruit, lizards, snakes, young birds, insects. **VOICE** loud cackles and screams; also a rapid *pak… pak…pak.* **DISTRIBUTION** Haryana and Uttarakhand to extreme NE; E Ghats, south to Bastar and N Andhra Pradesh. The Malabar Pied is absent in NE regions, but is found over the W Ghats. **HABITAT** forests, orchards, groves.

ABOVE: *Malabar Pied Hornbill*

Great Barbet ■ *Megalaima virens* 33cm

DESCRIPTION Sexes alike. Bluish-black head and throat; maroon-brown back; yellowish hind-collar; green on lower back and tail; brown upper breast; pale yellow below, with thick, greenish-blue streaks; red under tail-coverts distinctive. Large, yellowish beak. Either solitary or small bands; arboreal, but comes into low-fruiting bushes; difficult to spot and mostly heard. **FOOD** fruit, flower petals. **VOICE** very noisy, especially between March

and July; loud, if somewhat mournful *pi…you* or *pi… oo*, uttered continuously for several minutes; one of the most familiar bird calls in the Himalaya; often joined by the rather similar but high-pitched, more nasal calls of the **Golden-throated Barbet** M. *franklinii* (23cm), of the E Himalaya. **DISTRIBUTION** Himalaya, 800–3,200m. **HABITAT** forests, orchards.

RIGHT: *Golden-throated Barbet*

Brown-headed Barbet ▪ *Megalaima zeylanica* 28cm

DESCRIPTION Sexes alike. Grass-green plumage; brownish head, neck and upper back, streaked white; bare orange patch around eye. The **White-cheeked Barbet** M. *viridis* (23cm) of S India, has a white cheek-stripe. Solitary or in pairs; occasionally small parties; strictly arboreal; keeps to fruiting trees, often with other frugivorous birds; difficult to spot in the canopy; noisy in hot season; strong, undulating flight. **FOOD** chiefly fruits; also flower nectar, petals, insects and small lizards. **VOICE** noisy; its *kutroo...kutroo* or *pukrook... pukrook* calls one of the most familiar sounds of the Indian forests; calls often begin with a guttural *kurrrr*.

DISTRIBUTION most of India south of the Himalayan foothills (Himachal Pradesh to Nepal). **HABITAT** forests, groves; also city gardens.

White-cheeked Barbet

Blue-throated Barbet
▪ *Megalaima asiatica* 23cm

DESCRIPTION Sexes alike. Grass-green plumage; black, crimson, yellow and blue about head; blue chin and throat diagnostic; crimson spots on sides of throat. Solitary or in pairs; sometimes small parties on fruiting trees, along with other fruit-eating birds; strictly arboreal; keeps to canopy of tall trees; difficult to spot but loud, monotonous calls an indicator of its presence. **FOOD** chiefly fruits; also insects. **VOICE** calls similar to those of Brown-headed Barbet of the plains; on careful hearing, sounds somewhat softer and there is a short note between the two longer ones; can be interpreted as *kutt...oo...ruk...*; also a 4-note song when breeding. **DISTRIBUTION** the Himalaya east from Pakistan and Kashmir; found up to about 2,250m; also found in Bengal, including Kolkata. **HABITAT** forests, groves.

Coppersmith Barbet ▪ *Megalaima haemacephala* 17cm

DESCRIPTION Sexes alike. Grass-green plumage; yellow throat; crimson breast and forehead; dumpy appearance. The **Crimson-fronted Barbet** M. *rubricapillus* of the W Ghats, south of Goa, has a crimson chin, throat, foreneck and upper breast. Solitary, in pairs or small parties; strictly arboreal; feeds on fruiting trees, often with other birds; visits flowering *Erythrina* and *Bombax* trees for flower nectar; often spends early morning sunning itself on bare branches. **FOOD** chiefly fruits and berries; sometimes catches insects. **VOICE** noisy between December and end of April; monotonous *tuk… tuk…* calls one of the best-known bird calls of India, likened to a coppersmith working on his metal. **DISTRIBUTION** all India, up to about 1,800m in outer Himalaya. **HABITAT** light forests, groves, city gardens, roadside trees.

ABOVE: *Crimson-fronted Barbet*

Speckled Piculet ▪ *Picumnus innominatus* 10cm

DESCRIPTION Sexes alike. Olive-green above (male has some orange and black on forecrown); two white stripes on sides of head, the upper one longer; dark-olive band through eyes, moustachial stripe; creamy-white below, boldly spotted with black. Usually pairs; moves around thin branches, or clings upside-down; taps with beak, probes crevices; typical woodpecker behaviour; associates in mixed hunting bands; unobtrusive, hence often overlooked; perches across branches. **FOOD** chiefly ants and termites. **VOICE** sharp, rapid *tsip…tsip…*; also a loud drumming sound. **DISTRIBUTION** Himalaya, west to east, foothills to at least 2,500m. The slightly duller southern race *malayorum* has a wide distribution over the E Ghats and the W Ghats, south of Goa; also Nilgiris, Palnis and associated mountain ranges. **HABITAT** mixed forests, with a fondness for bamboo jungle.

Brown-capped Pygmy Woodpecker ■ *Dendrocopos nanus* 13cm

ABOVE: *Grey-capped Pygmy Woodpecker*

DESCRIPTION Small woodpecker. Male: barred brown and white above; paler crown with short, scarlet streak (occipital); prominent white band from just above eyes extends to neck; pale dirty-brown-white below, streaked black. Female: like male but lacks the scarlet streaks on sides of crown. The male **Grey-capped Pygmy Woodpecker** *D. canicapillus* (14cm) of the Himalaya has short scarlet occipital crest; black upper back and white-barred lower back and rump. Mostly in pairs; often a part of mixed bird parties in forest; seen more on smaller trees, branches and twigs, close to ground and also high in canopy; quite active. **FOOD** small insects, grubs, obtained from crevices and under bark; also small berries. **VOICE** faint but shrill squeak, sounds like *clicck…rrr*. **DISTRIBUTION** almost all over India, including some of the drier regions of N India. **HABITAT** light forests, cultivation, bamboos, orchards; also vicinity of habitation.

Streak-throated Woodpecker

■ *Picus xanthopygaeus* 30cm

DESCRIPTION Male: grass-green above; crimson crown and crest; orange and black on nape; white supercilium and malar stripe; yellow rump; bold, black scaly streaks on whitish underbody, with tawny-green wash on breast; throat greyer, also streaked. Female: black crown and crest. Solitary or in pairs; works up along tree stems; moves either straight up or in spirals; taps with beak for insects hiding in bark; also settles on ground. **FOOD** mostly insects: ants, termites, wood-boring beetle larvae; also figs. **VOICE** occasional faint *pick…* mostly silent; also drums on branches. **DISTRIBUTION** all subcontinent; found up to 1,500m in outer Himalaya. **HABITAT** mixed forests, plantations.

Himalayan Woodpecker
■ *Dendrocopos himalayensis* 25cm

DESCRIPTION Male: black back and upper body; white shoulder-patch; white spots and barring on wings; crimson crown and crest; white lores, cheeks and ear-coverts; broad black moustachial stripe; yellowish-brown underbody, darker on breast; crimson under tail. Female: black crown and crest. Mostly in pairs, moving about in forest; jerkily moves up and around tree stems or clings on undersides of branches; like other woodpeckers often moves a few steps back, as if to re-examine; sometimes seen in mixed hunting parties of Himalayan birds. **FOOD** mostly insects hunted from under the bark and moss; seeds of conifers; nuts and acorns. **VOICE** fairly loud calls, uttered in night. **DISTRIBUTION** Himalaya, from Kashmir to W Nepal; 1,500–3,200m. **HABITAT** Himalayan forests.

Rufous Woodpecker ■ *Micropternus brachyurus* 25cm

DESCRIPTION Sexes alike. Chestnut-brown plumage; fine black crossbars on upper body, including wings and tail; paler edges to throat feathers; crimson patch under eye in male,

absent in female. Usually in pairs; sometimes 4 or 5 scattered birds close by; mostly seen around ball-shaped nests of tree ants; clings to outsides of nests and digs for ants; plumage often smeared with gummy substance. **FOOD** chiefly tree ants and their pupae; occasionally figs and other fruit; seen to suck sap from near bases of banana leaves.

VOICE rather vocal between January and April; loud, high-pitched 3 or 4 notes *ke…ke… kr…ke…* drums when breeding. **DISTRIBUTION** subcontinent, south of outer Himalaya, found up to 1,500m. **HABITAT** mixed forests.

Yellow-crowned Woodpecker
■ *Dendrocopos mahrattensis* 18cm

DESCRIPTION Male: brownish-black above, spotted all over with white; golden-brown forehead and crown; small scarlet crest; pale fulvous below throat, streaked brown; scarlet patch in centre of abdomen distinctive. Female: lacks scarlet crest. Solitary or pairs, sometimes small bands of up to 6 birds; occasionally seen with mixed hunting parties; moves in jerks along tree stems and branches; hunts in typical woodpecker manner; rather confiding in some areas; birds keep in touch with faint creaking sounds. **FOOD** chiefly insects; also figs, other fruits and flower nectar. **VOICE** soft but sharp *clic…click…clickrrr…*; drums when breeding. **DISTRIBUTION** common and widespread; almost subcontinent, from Himalayan foothills south; uncommon in NE regions. **HABITAT** open forests, scrub, cultivation, vicinity of habitation, gardens.

Heart-spotted Woodpecker
■ *Hemicircus canente* 16cm

DESCRIPTION Male: black forehead (speckled white), crown and crest; black back; broad, pale buff wing-patch (inner secondaries and wing-coverts) with heart-shaped spots; black flight feathers; whitish-buff, olive and black below. Female: extensive buff white on forehead, otherwise like male. Pairs or small parties; active and arboreal; perches across branches and calls often as it flies from one tree to another. **FOOD** insects, mostly ants and termites. **VOICE** quite vocal, especially in flight; a somewhat harsh *chur…* note; other sharp clicking and squeaky notes. **DISTRIBUTION** W Ghats from Kerala north to Tapti river; east across Satpuras to SE Madhya Pradesh, Orissa, NE states. **HABITAT** forests.

Grey-headed Woodpecker
■ *Picus canus* 32cm

DESCRIPTION Male: darkish green above; crimson forehead; black hind-crown, faint crest and nape; dark sides of head and black malar stripe; yellow rump, white-barred dark wings and blackish tail; unmarked, dull greyish-olive underbody diagnostic. Female: black from forehead to nape; no crimson. Solitary or in pairs; typical woodpecker, moving on tree stems and larger branches, hunting out insects from under bark; descends to ground, hopping awkwardly; also digs into termite mounds. **FOOD** termites, ants, wood-boring beetles and their larvae; also feeds on flower nectar and fruits. **VOICE** loud, chattering alarm; common call is a high-pitched *keek…keek…* of 4 or 5 notes; drums often between March and early June. **DISTRIBUTION** Himalaya from the lower foothills country to about 2,700m. **HABITAT** forests, both deciduous and temperate.

Lesser Goldenback
■ *Dinopium benghalense* 30cm

DESCRIPTION Male: shining golden-yellow and black above; crimson crown and crest; black throat and sides of head, with fine white streaks; white underbody, streaked black, boldly on breast. Female: black crown spotted with white; crimson crest. Usually pairs, sometimes half a dozen together; widespread and common; moves jerkily up and around tree stems or clings on undersides of branches; taps out insects; often associates in mixed hunting parties; may descend to ground, picking off ants and other insects. **FOOD** chiefly ants, termites; caterpillars and centipedes on ground; also figs and berries. **VOICE** noisy; loud, high-pitched cackle, like laughter; drums often. **DISTRIBUTION** subcontinent, up to about 1,800m in outer Himalaya; also found in drier areas of NW India. **HABITAT** forests, both dry and mixed deciduous; orchards; gardens; also neighbourhood of villages and other habitation.

Greater Goldenback ■ *Chrysocolaptes lucidus* 32cm

DESCRIPTION Male: crimson crown and crest; golden-olive above; white and black sides of face and throat; whitish-buff below, profusely spotted with black on foreneck, and speckled over rest of underbody; extensive crimson rump and black tail and flight feathers distinctive. Female: white-spotted black crown and crest. The **Himalayan Goldenback** *D. shorii* is very similar, but slightly smaller size, black nape, three toes and two narrow stripes down throat centre can help make the distinction. The **Common Goldenback** *D. javanense* (28cm) is also confusingly similar, but has single black malar stripe. Pairs or small bands; arboreal; moves jerkily up along tree stems. **FOOD**

insects; possibly nectar. **VOICE** noisy; loud, grating scream; calls mostly in flight. **DISTRIBUTION** Uttarakhand to NE; parts of E Ghats, SE Madhya Pradesh; W Ghats, Kerala to Tapti river; plains to about 1,500m. **HABITAT** forests.

Common Goldenback   *Himalayan Goldenback*

Indian Pitta ■ *Pitta brachyura* 19cm

DESCRIPTION Sexes alike. A multi-coloured, stub-tailed, stoutly built bird; bright blue, green, black, white, yellowish-brown and crimson; white chin, throat and patch on wing-tips and crimson vent distinctive. Solitary or pairs; small flocks on migration, before and after monsoons; spends much time on ground, hopping about, hunting for insects amidst the leaf litter and low herbage; quietly flies into a tree branch if disturbed; shows fondness for shaded, semi-damp areas. **FOOD** chiefly insects. **VOICE** loud, lively whistle, *wheeet...peu*; very vocal when breeding (during rains); also a longish single-note whistle. **DISTRIBUTION** almost entire subcontinent, with considerable seasonal movement, particularly before and after the rains; breeds commonly over N and C India; also elsewhere. **HABITAT** forests, orchards; also cultivated country.

Common Woodshrike ■ *Tephrodornis pondicerianus* 16cm

DESCRIPTION Sexes alike. Greyish-brown plumage; broad whitish supercilium and dark stripe below eye distinctive; white outer tail feathers seen when bird flies. Dark stripe may

be slightly paler in female. The **Malabar Wood Shrike** *T. v. sylvicola* is larger (23cm) and has white outer tail feathers. Pairs or small parties; quiet for greater part of year, vocal when breeding (February–May); keeps to middle levels of trees, hopping about, sometimes coming to ground. **FOOD** insects; also flower nectar. **VOICE**

whistling *wheet...wheet...* and an interrogative, quick-repeated *whi...whi...whi...whee* thereafter; other trilling, pleasant notes when breeding. **DISTRIBUTION** most of the country, south of Himalayan foothills; most common in low country. **HABITAT** light forests, edges of forest, cultivation, gardens in and around habitation.

LEFT: *Malabar Wood Shrike*

Common Iora ■ *Aegithina tiphia* 14cm

DESCRIPTION Male: greenish above (rich black above, with yellowish rump, in summer breeding plumage); black wings and tail; two white wing-bars; bright yellow underbody. Female: yellow-green plumage; white wing-bars; greenish-brown wings. Pairs keep to leafy branches, often with other small birds; moves energetically amidst branches in hunt for insects, caterpillars; rich call notes often a giveaway of its presence in an area. **FOOD** insects, spiders; also flower nectar. **VOICE** renowned vocalist; wide range of rich, whistling notes; single or 2-note long-drawn *wheeeeeee* or *wheeeeeee... chu* is a common call; another common call is a 3-note whistle. **DISTRIBUTION** subcontinent, up to about 1,800m in the Himalaya; absent in arid NW, desert regions of Rajasthan, Kutch. **HABITAT** forest, gardens, orchards, tree-dotted cultivation, habitation.

Small Minivet ■ *Pericrocotus cinnamomeus* 15cm

DESCRIPTION Male: dark grey head, back and throat; orange-yellow patch on black wings; black tail; flame-orange breast; orange-yellow belly and undertail. Female: paler above; orange rump; dusky white throat, breast tinged with yellow; yellowish belly and under tail. Pairs or small flocks; keep to tree-tops, actively moving amidst foliage; flutters and flits about in an untiring hunts for small insects, often in association with other small

birds; also hunt flycatcher style. **FOOD** chiefly insects; sometimes flower nectar. **VOICE** soft, low *swee...svee...* notes uttered as birds hunt in foliage. **DISTRIBUTION** most of India, up to about 900m in outer Himalaya; absent in arid parts of Rajasthan. **HABITAT** forests, groves, gardens, tree-dotted cultivation.

Scarlet Minivet ■ *Pericrocotus (flammeus) speciosus* 20cm

DESCRIPTION Male: glistening black head and upper back; deep scarlet lower back and rump; black and scarlet wings and tail; black throat, scarlet below. Female: rich yellow forehead, supercilium; grey-yellow above; yellow and black wings and tail; bright yellow underbody. Pairs or small parties; sometimes several dozen together; keeps to canopy of

tall trees; actively flits about to hunt for insects; also launches aerial sallies after winged insects; often seen in mixed hunting parties of birds; spectacular sight of black, scarlet and yellow as flock flies over forest, especially when seen from above. **FOOD** insects, flower nectar. **VOICE** pleasant, 2-note

whistle; also a longer, whistling warble. **DISTRIBUTION** disjunct; several isolated races. The recently split **Orange Minivet** *P. flammeus* is found south of Gujarat through the W Ghats. **HABITAT** forests, gardens, groves.

Brown Shrike
■ *Lanius cristatus* 19cm

DESCRIPTION Uniformly rufous-brown upperparts; black band through the eye with a white brow over it. Pale creamy underside with warmer rufous flanks; rufous tail. Wings brown without any white 'mirror'. Female has faint scalloping on the underside. Solitary; keeps lookout from conspicuous perch or tree stump for prey on the ground, often returning to the same perch after hunting; territorial. **FOOD** insects, lizards, small rodents. **VOICE** harsh chattering; grating call; sometimes sings in a low chirruping tone with its bill closed. **DISTRIBUTION** winter visitor to peninsular India. **HABITAT** open country, cultivation, forest edges, scrub, gardens.

Bay-backed Shrike
■ *Lanius vittatus* 18cm

DESCRIPTION Sexes alike. Deep chestnut-maroon back; broad black forehead-band, continuing through eyes to ear-coverts; grey crown and neck, separated from black by small white patch; white rump distinctive; black wings with white in outer flight feathers; white underbody, fulvous on breast and flanks. Solitary or in scattered pairs in open terrain; keeps lookout from a perch on some tree stump, overhead wire or bush top, usually under 4m off ground; pounces once potential prey is sighted; usually devours prey on ground, tearing it; sometimes carries it to perch; keeps to fixed territories, defended aggressively. **FOOD** insects, lizards, small rodents. **VOICE** harsh *churr*; lively warble of breeding male; sometimes imitates other bird calls. **DISTRIBUTION** subcontinent, up to about 1,800m in the Himalaya; absent in the NE. **HABITAT** open country, light forests, scrub.

Long-tailed Shrike ▪ *Lanius schach* 25cm

DESCRIPTION Sexes alike. Pale grey from crown to middle of back; bright rufous from then on to the rump; black forehead, band through eye; white 'mirror' in black wings; whitish underbody, tinged pale rufous on lower breast and flanks. Mostly solitary; boldly defends feeding territory; keeps lookout from conspicuous perch; pounces onto ground on sighting prey; said to store surplus in 'larder', impaling prey on thorns; nicknamed Butcher-bird. **FOOD** insects, lizards, small rodents, birds. **VOICE** noisy; harsh mix of scolding notes, shrieks and yelps; excellent mimic; rather musical song of breeding male. **DISTRIBUTION** three races; undergo considerable seasonal movement; subcontinent, from about 2,700m in the Himalaya. The black-headed subspecies *L. s. tricolor*, has a black head and small white patch on wings; breeds in the Himalaya east of Uttarakhand. **HABITAT** open country, cultivation, edges of forest, vicinity of habitation, gardens; prefers neighbourhood of water.

L. s. tricolor

Southern Grey Shrike
▪ *Lanius meridionalis* 25cm

DESCRIPTION Sexes alike. Bluish-grey above; broad black stripe from beak through eye; black wings with white mirrors; black and white tail; unmarked white underbody. Mostly in pairs in open areas; remains perched upright on bush tops or overhead wires or flies low, uttering a harsh scream; surveys neighbourhood from perch and pounces on prey; batters and tears prey before swallowing it; said to maintain a larder, impaling surplus prey on thorns; keeps feeding territories year round; a wild and wary bird. **FOOD** insects, lizards, small birds, rodents. **VOICE** harsh, grating *khreck...* call; a mix of other harsh notes and chuckles; pleasant, ringing song of breeding male. **DISTRIBUTION** the drier areas of NW, W and C India, across Gangetic plain to W Bengal; south to Tamil Nadu. **HABITAT** open country, semi-desert, scrub, edges of cultivation.

Indian Golden Oriole ■ *Oriolus (oriolus) kundoo* 25cm

DESCRIPTION Male: bright golden-yellow plumage; black stripe through eye; black wings and centre of tail. Female: yellow-green above; brownish-green wings; dirty-white below, streaked brown. Young male much like female. Solitary or in pairs; arboreal, sometimes

moving with other birds in upper branches; regularly visits fruiting and flowering trees; hunts insects in leafy branches; usually heard, surprisingly not often seen, despite bright colour; seen only when it emerges on bare branch or flies across. **FOOD** insects, fruit, nectar. **VOICE** fluty whistle of 2 or 3 notes, interpreted *pee…lo…lo*, the middle note lower; harsh note often heard; rich, mellow song when breeding, somewhat mournful; does not sing often. **DISTRIBUTION** summer visitor to the Himalayan foothills to about 2,600m; spreads in winter to plains; breeds also in many parts of peninsula. **HABITAT** forest, orchards, gardens around habitation.

Black-hooded Oriole
■ *Orialus xanthornus* 25cm

DESCRIPTION Sexes alike. Golden-yellow plumage; black head diagnostic; black and yellow wings and tail; deep pink-red beak seen at close quarters. Pairs or small parties; strictly arboreal, only rarely descending into lower bushes or to ground; active and lively; moves a lot in forest and birds chase one another, the rich colours striking against green or brown of forest; very vocal; associates with other birds in mixed parties; visits fruiting and flowering trees. **FOOD** fruits, flower nectar, insects. **VOICE** assortment of melodious and harsh calls; more common is a fluty 2- or 3-noted *tu…hee* or *tll…yow…yow…*; also a single, mellow note. **DISTRIBUTION** subcontinent, up to about 1,000m in the Himalayan foothills. **HABITAT** forests, orchards, gardens, often amidst habitation.

Black Drongo ■ *Dicrurus macrocercus* 28cm

DESCRIPTION Sexes alike. Glossy black plumage; long, deeply forked tail. Diagnostic white spot at base of bill. The **Ashy Drongo** *D. leucophaeus* (30cm) is grey-black, and more of a forest bird, breeding in Himalaya and a winter visitor to the peninsula. Usually solitary, sometimes small parties; keeps lookout from exposed perch; most common bird seen on rail and road travel in India; drops to ground to capture prey; launches short aerial sallies; rides atop grazing cattle; follows cattle, tractors, grass-cutters, fires; thus consumes vast numbers of insects; bold and aggressive species, with several birds nesting in same tree. **FOOD** chiefly insects; supplements with flower nectar, small lizards. **VOICE** harsh *tiu-tiu* also *cheece cheece*. **DISTRIBUTION** subcontinent, up to about 1,800m in outer Himalaya. **HABITAT** open country, orchards, cultivation.

Ashy Drongo

Spangled Drongo
■ *Dicrurus hottentottus* 30cm

DESCRIPTION Sexes alike. Glistening blue-black plumage, fine hair-like feathers on forehead; longish, down-curved, pointed beak; diagnostic tail, square-cut and inwardly bent (curling) towards outer ends. Solitary or scattered pairs, strictly arboreal forest bird; small numbers may gather on favourite flowering trees; rather aggressive; often seen in mixed hunting parties of birds. **FOOD** chiefly flower nectar; also insects, more so when there are young in nest **VOICE** noisy; a mix of whistling, metallic calls and harsh screams. **DISTRIBUTION** lower Himalaya foothills, east of Uttarakhand, down through NE India, along E Ghats, Orissa, Bastar through to W Ghats, up north to Mumbai, occasionally even further north. **HABITAT** forests.

Greater Racket-tailed Drongo ■ *Dicrurus paradiseus* 60cm

DESCRIPTION Sexes alike. Glossy blue-black plumage; prominent crest of longish feathers, curving backwards; elongated, wire-like outer tail feathers, ending in 'rackets' diagnostic. Solitary or in pairs, sometimes small gatherings; arboreal forest bird, but often descends into low bush; moves a lot in forest; confirmed exhibitionist, both by sight and sound; extremely noisy, often vocal long before sunrise;

bold and aggressive, seen mobbing bigger birds 100m over forest. The **Lesser Racket-tailed Drongo** *D. remifer* (38cm) is found in lower Himalaya, east of Uttarakhand. **FOOD** mostly insects; also lizards, flower nectar. **VOICE** noisiest bird of forest; amazing mimic; wide variety of whistles and screams, perfect imitations of over a dozen species. **DISTRIBUTION** forested parts of India, roughly east and south of line from S Gujarat to Uttarakhand; up to about 1,400m.

ABOVE RIGHT: *Lesser Racket-tailed Drongo* **HABITAT** forests; also forest edges, orchards.

White-throated Fantail ■ *Rhipidura albicollis* 17cm

DESCRIPTION Sexes alike. Slaty-brown plumage, including underbody; short, white supercilium; white throat and tips to all but central tail feathers. The **White-spotted Fantail** *R. albogularis* have a white-spotted slaty band across breast; also, whitish-buff belly, less white in tail-tip. Overall behaviour not appreciably different from **White-browed Fantail** *R. aureola*. **FOOD** insects, spiders. **VOICE** a harsh *chukrrr…* note. **DISTRIBUTION** outer Himalaya, to about 1,800m; NE regions; absent through Indo-Gangetic plain west to NW parts of India; two races from Orissa to Godavari river; most widespread is the White-spotted race *albogularis*, found all over peninsular India, south of a line from S Rajasthan, across the Vindhya and south along the edge of E Ghats. **HABITAT** light forests, groves, gardens amidst habitation, scrub.

ABOVE: *White-spotted Fantail*   *White-browed Fantail*

Black-naped Monarch
■ *Hypothymis azurea* 16cm

DESCRIPTION Male: lilac-blue plumage;
black patch on nape, gorget on breast; slight
black scaly markings on crown; sooty on wings
and tail; white below breast. Female: ashy-blue,
duller; lacks black on nape and breast. Solitary
or pairs in forest, often amidst mixed hunting
parties; extremely active and fidgety, flits and
flutters about, often fans tail slightly; calls
often as it moves about, the calls often the
first indication of its presence. **FOOD** insects.
VOICE common call a sharp, grating, high-
pitched *chwich...chweech* or *chwae...chweech*,
slightly interrogative in tone, the two notes
quickly uttered; has short, rambling notes
when breeding. **DISTRIBUTION** India south
of outer Himalaya, to about 1,200m, east of
W Uttarakhand; absent in arid NW, and
N India. **HABITAT** forest, bamboo, gardens.

Asian Paradise-flycatcher
■ *Terpsiphone paradisi* Adult male: 50cm including tail-streamers.

DESCRIPTION Glossy blue-black head, crest and throat; black
in wings; silvery-white body, long tail-streamers. In rufous phase
white parts replaced by rufous chestnut. Female and young male:
20cm. No tail-streamers; shorter crest; rufous above; ashy-grey
throat and nuchal collar; whitish below.
Solitary or pairs; makes short sallies; flits
through trees, the tail-streamers floating;
strictly arboreal, sometimes descending
into taller bushes; cheerful disposition.
FOOD insects, spiders. **VOICE** sharp,
grating *chwae* or *chchwae...* call;
melodious warbling song and display
of breeding male. **DISTRIBUTION**
Himalaya, foothills to about 1,800m,
rarely 2,500m; N India, south to
Bharatpur; absent in a broad belt across
Gangetic plain; widespread in peninsular
India. **HABITAT** light forests, gardens,
open country.

Eurasian Jay ■ *Garrulus glandarius* 33cm

DESCRIPTION Sexes alike. Pinkish-brown plumage; velvet-black malar stripe; closely black-barred, blue wings; white rump contrasts with jet-black tail. The **Black-headed Jay** G. *lanceolatus* of the W Himalaya, east to about C Nepal, has a black cap, black and white face, and white in wings. Small, noisy bands,

often along with other Himalayan birds; common and familiar about Himalayan hill stations; inquisitive and aggressive; mostly keeps to trees, but also descends into bush and on to ground; laboured flight. **FOOD** insects, fruits, nuts. **VOICE** noisy; guttural chuckles, screeching notes and whistles; good mimic. **DISTRIBUTION** across Himalaya, 1,500–2,800m, somewhat higher in the east; may descend low in winter. **HABITAT** mixed temperate forests.

ABOVE: *Black-headed Jay*

Yellow-billed Blue Magpie ■ *Urocissa flavirostris* 66cm

DESCRIPTION Sexes alike. Purple-blue plumage; black head and breast; white nape-patch and underbody; very long, white-tipped tail; yellow beak and orange legs. The **Red-billed Blue Magpie** U. *erythrorhyncha* (70cm) has more white on nape; red beak and legs; appears to be restricted to the Himalaya between Himachal Pradesh and E Nepal and some parts of north-east India. Pairs or small bands, often associating with jays, laughingthrushes and treepies; wanders a lot in forests, flying across clearings, entering hill station gardens, one bird following another; arboreal, but also hunt low in bushes; even descends to ground, the long tail cocked as the bird hops about. **FOOD** insects, fruit, lizards, eggs, small birds. **VOICE** noisy; great mix of metallic screams, loud whistles and raucous notes, often imitating other birds. **DISTRIBUTION** Himalaya, west to east; 1,500–3,600m; may descend low in winter. **HABITAT** forests, gardens, clearings.

LEFT: *Red-billed Blue Magpie*

Rufous Treepie ▪ *Dendrocitta vagabunda* 50cm

DESCRIPTION Sexes alike. Rufous above; sooty grey-brown head and neck; black, white and grey on wings, best seen in flight; black-tipped, grey tail, long and graduated. Pairs or small parties; often seen in mixed hunting parties, appearing as leader of pack; feeds up in trees, but also descends low into bushes and onto ground to pick up termites; bold and noisy; rather tame and confiding in certain areas. The **Grey Treepie** *D. formosae* replaces the Rufous Treepie at higher altitudes. **FOOD** insects, lizards, small birds, eggs, fruit, flower nectar; kitchen scraps in some areas. **VOICE** common call a fluty 3-note *goo…ge…lay* or *ko… ki…la*; harsh, guttural notes often uttered. **DISTRIBUTION** almost all India, up to about 1,500m in outer Himalaya. **HABITAT** forests, gardens, cultivation, habitation.

Grey Treepie

House Crow ▪ *Corvus splendens* 43cm

DESCRIPTION Sexes alike. Black plumage; grey collar, upper back and breast; glossy black on forehead, crown and throat. The **Eurasian Jackdaw** *C. monedula* (33cm) is similar to the House Crow, but is smaller, thicker-necked and white-eyed; it is common in Kashmir. Described as an extension of man's society; street-smart, sharp, swift, sociable, sinister, the crow is almost totally commensal on man, snatches food from tables and shops; mobs other birds, even large raptors; performs important scavenging services; occasionally flies very high into skies, either when flying long distance, or simply for fun, communal roost sites. **FOOD** omnivorous; robs young birds from nests; drives other birds from flowering trees. **VOICE** familiar *caw* call; occasionally a pleasant *kurrrrr* note; several other notes. **DISTRIBUTION** subcontinent, reaching about 2,500m in the Himalaya. **HABITAT** habitation rural and urban; cultivation, forest edges; range of habitats very wide.

Eurasian Jackdaw

Large-billed Crow ■ *Corvus macrorhynchos* 48cm

DESCRIPTION Sexes alike. Glossy black plumage; heavy beak, with noticeable culmen-curve. The **Carrion Crow** C. *corone* of NW mountains is confusingly similar, except for less curved culmen though this character not easily visible in the field. Solitary or in groups of two to six; most common around villages and only small numbers in urban areas; overall not as 'enterprising' as the familiar House Crow; in forested areas, its behaviour often indicates presence of carnivore kills. **FOOD** omnivorous. **VOICE** harsh *khaa…khaa* calls; several variations on this among the various races of this crow. **DISTRIBUTION** subcontinent, to about 4,500m in the Himalaya; absent in extreme W Rajasthan and parts of Punjab. **HABITAT** forests, rural habitations; small numbers in towns and cities.

ABOVE: *Carrion Crow*

Grey-headed Canary Flycatcher ■ *Culicicapa ceylonensis* 9cm

DESCRIPTION Sexes alike. Ashy-grey head, throat and breast; darker crown; yellow-green back and yellow rump; yellow in browner wings and tail; yellow below breast. Some flycatcher-warblers (*Seicercus* spp) are superficially similar to the Grey-headed Canary Flycatcher, but lack grey on throat and breast. Solitary or in pairs, occasionally several in vicinity, especially in mixed parties; a forest bird, typical flycatcher, excitedly flitting about, launching aerial sallies and generally on the move; wherever this bird is, its cheerful unmistakable calls are heard. **FOOD** insects. **VOICE** vocal; high-pitched 2- or 3- syllabled calls, *whi…chichee…whi…chichee*; longer, trilling song; also chattering notes. **DISTRIBUTION** commonly breeds in the Himalaya, 1,500–3,000m; possibly in some of the hill forests of C India and E Ghats; common in winter over much of the subcontinent. **HABITAT** forests, gardens, orchards.

Great Tit ■ *Parus major* 13cm

DESCRIPTION Sexes alike. Grey back; black crown continued along sides of neck to broad black band from chin along centre of underbody; white cheeks, nape-patch, wing bar and outer feathers of black tail; ashy-white sides. The **White-naped Tit** *P. nuchalis* of W India lacks black on neck sides; has extensive white in wings and sides of body. Pairs or small bands, often with other small birds; restless, clings upside down, and indulges in all sorts of acrobatic displays as it hunts amongst leaves and branches; holds food fast between feet and pecks at it noisily; tame and confiding. **FOOD** insects, small fruit. **VOICE** loud, clear whistling *whee…chi…chee…*; other whistling and harsh notes. **DISTRIBUTION** widespread in

Himalaya, foothills to about 3,500m; peninsular India from Gujarat, C Rajasthan and Orissa south; absent in broad belt from NW India across Gangetic plains. **HABITAT** open forests, gardens, habitation.

White-naped Tit

Black-lored Tit ■ *Parus xanthogenys* 14cm

DESCRIPTION Sexes alike. Olive-green back; black crest (faintly tipped with yellow), stripe behind eye, broad central band from chin to vent; bright yellow nape-patch, supercilium and sides of underbody. The **Yellow-cheeked Tit** *P. spilonotus* has black streaks on back, and yellow lores and forehead. Pairs or small flocks, often with other small birds; arboreal, active; feeds in foliage; sometimes enters gardens. **FOOD** insects, berries. **VOICE** cheerful, musical notes; loud tailorbird like *towit…towit* calls near nest; other 2- to 4-noted whistling calls; whistling song; also harsh *charrr* and some chattering notes. **DISTRIBUTION** the Himalaya to E Nepal; 1,200–2,500m, widespread in parts of C, E and W India. **HABITAT** forests, gardens.

Yellow-cheeked Tit

Sultan Tit ▪ *Melanochlora sultanea* 20cm

DESCRIPTION Male: black above; yellow crown and crest; black throat and upper breast; yellow below. Female: deep olivish wash to black upper body and throat; crest as in male; some yellow also on throat. Small bands, often along with other birds in mixed hunting flocks; active and inquisitive; clings sideways and upside down; checks foliage and bark crevices; feeds in canopy but also descends to tall bushes. **FOOD** insects, small fruit, seeds. **VOICE** noisy; loud, whistling *cheerie...cheerie*; other shrill whistling notes, often mixed with a harsh *churr* or *chrrchuk*; varied chattering notes. **DISTRIBUTION** the Himalayan foothills, east from C Nepal; NE; foothills to about 1,200m, sometimes ascending to 2,000m. **HABITAT** mixed forests, edges of forest.

Indian Bushlark
▪ *Mirafra erythroptera* 14cm

DESCRIPTION Sexes alike. Yellowish-brown above, streaked black; rich chestnut-rufous on wings, easily seen when bird in flight: pale white chin and throat, dull yellowish-brown below; blackish, triangular spots on breast. Pairs or small flocks; moves quietly on ground, running about or perching on small stones or bush tops; squats tight when approached but takes to wing when intruder very close; spectacular display flight, accompanied by singing, when breeding; indulges in display flights in the night too. **FOOD** seeds, tiny insects. **VOICE** faint *cheep chrep* call note; song a faint but lively twittering. **DISTRIBUTION** almost all of N, NW and peninsular India; absent in Kerala, NE. **HABITAT** open cultivation, grass and scrub; fallow lands.

Crested Lark ▪ *Galerida cristata* 18cm

DESCRIPTION Sexes alike. Sandy-brown above, streaked blackish; pointed, upstanding crest distinctive; brown tail has dull rufous outer feathers; whitish and dull yellowish-brown below, the breast streaked dark brown. The **Malabar** G. *malabarica* (15cm) and **Sykes's Crested Larks** G. *deva* (13cm) are very similar, but overall plumage is darker, more rufous-brown; also both are birds of peninsular and S India. Small flocks, breaking into pairs when breeding; runs briskly on ground, the pointed crest carried upstanding; also settles on bush tops, stumps, wire fences, overhead wires. **FOOD** seeds, grain, insects. **VOICE** ordinary call note a pleasant *tee...ur...*; short song of male during soaring display flight. **DISTRIBUTION** N, NW India, Gangetic plain; Rajasthan, Saurashtra, N Madhya Pradesh. **HABITAT** semi-desert, cultivation, dry grassy areas.

Malabar Lark

Syke's Crested Lark

Ashy-crowned Sparrow Lark ▪ *Eremopterix griseus* 13cm

DESCRIPTION Thickish beak. Male: sandy-brown above; white cheeks and sides of breast; dark chocolate-brown sides of face and most of underbody; dark brown tail with whitish outer feathers. Female: sandy-brown overall; dull rufous sides of face and underbody. Mostly loose flocks, scattered over an area; pairs or small parties when breeding; feed on ground; fond of dusty areas, where large numbers may squat about; sandy colouration makes it impossible to spot the birds, but when disturbed, large numbers suddenly take wing; superb display flight of male. **FOOD** grass seeds, tiny insects. **VOICE** pleasant, monotonous trilling song by male; sings on wing and on ground. **DISTRIBUTION** almost entire India, south of Himalayan foothills; moves during the rains; uncommon in heavy rainfall areas. **HABITAT** open scrub, semi cultivation, fallow river basins, tidal mudflats.

Black-crested Bulbul ■ *Pycnonotus m. flaviventris* 15cm

DESCRIPTION Sexes alike. Glossy black head, crest and throat; olive-yellow nape and back, becoming brown on tail; yellow below throat. The **Flame-throated Bulbul** *P. m. gularis* of W Ghats has a ruby-red throat. Pairs or small bands, sometimes with other birds; arboreal. **FOOD** insects, fruit. **VOICE** cheerful whistles; also a harsh *cburr* call; 4- to 8-note song. **DISTRIBUTION** Himalaya, from Himachal Pradesh eastwards; NE; foothills to about 2,000m. **HABITAT** forests, bamboo, clearings, orchards.

ABOVE: *Flame-throated Bulbul*

Red-whiskered Bulbul

■ *Pycnonotus jocosus* 20cm

DESCRIPTION Sexes alike. Brown above, slightly darker on wings and tail; black perky crest distinctive; crimson 'whiskers' behind eyes; white underbody with broken breast-collar; crimson-scarlet vent. Sociable; pairs or small flocks, occasionally gatherings of up to 100 birds; lively and energetic; feeds in canopy, low bush and on ground; enliven its surroundings with cheerful whistling notes; tame and confiding in some areas. **FOOD** insects, fruits, flower nectar. **VOICE** cheerful whistling notes; also harsh, grating alarm notes. **DISTRIBUTION** from Uttarakhand east along Himalayan foothills to about 1,500m; most common south of Satpura mountains in peninsular India; disjunct population in hilly areas of S, SE Rajasthan and N Gujarat. **HABITAT** forests, clearings, gardens and orchards, vicinity of human habitation.

Red-vented Bulbul
▪ *Pycnonotus cafer* 20cm

DESCRIPTION Sexes alike. Dark sooty-brown plumage; pale edges of feathers on back and breast give scaly appearance; darker head, with slight crest; almost black on throat; white rump and red vent distinctive; dark tail tipped white. Pairs or small flocks, but large numbers gather to feed; arboreal, keeps to middle levels of trees and bushes; a well known Indian bird, rather attached to man's neighbourhood; pleasantly noisy and cheerful, lively and quarrelsome; indulges in dust-bathing; also hunts flycatcher-style. **FOOD** insects, fruits, flower nectar, kitchen scraps. **VOICE** cheerful whistling calls; alarm calls on sighting snake, owl or some other intrusion, serving to alert other birds. **DISTRIBUTION** subcontinent, to about 1,800m in Himalaya. **HABITAT** light forests, gardens, haunts of man.

White-browed Bulbul ▪ *Pycnonotus luteolus* 20cm

DESCRIPTION Sexes alike. Olive plumage, brighter above; whitish forehead and supercilium, and explosive calls confirm identity. Pairs or small parties; not an easy bird to see; skulks in dense, low growth, from where its chattering calls suddenly explode; seen only momentarily when it emerges on bush tops, or flies low from one bush patch to another; usually does not associate with other birds. **FOOD** insects, fruits, nectar. **VOICE** loud, explosive chatter, an assortment of bubbling, whistling notes and chuckles. **DISTRIBUTION** peninsular India, south of a line from C Gujarat to S West Bengal; avoids the heavy rainfall hill zones of W Ghats. **HABITAT** dry scrub, village habitation, light forests, clearings.

Himalayan Bulbul ■ *Pycnonotus leucogenys* 20cm

DESCRIPTION Sexes alike. Brown head, a front-pointed crest and a short, white superciliary stripe. Black around eyes and on chin and throat. The **White-eared Bulbul** *P. leucotis* of the plains is light-brown above with a black head and throat; white cheeks: dark brown tail, tipped white; yellow vent. Pairs or small parties; active birds on the move, attracting attention by their pleasant calls; the Himalayan bird is common in the hills, where it is quite confiding. **FOOD** insects, fruits, flower nectar. **VOICE** pleasant whistling notes. **DISTRIBUTION** found in the Himalaya, from the foothills to about 3,400m. **HABITAT** open scrub, vicinity of habitation, edges of forest.

ABOVE: *White-eared Bulbul*

Black Bulbul ■ *Hypsipetes leucocephalus* 23cm

DESCRIPTION Sexes alike. Ashy-grey plumage; black, loose-looking crest; coral-red beak and legs diagnostic; whitish below the abdomen. Flocks in forest, often dozens together; strictly arboreal, keeps to topmost branches of tall forest trees, rarely comes down into undergrowth; noisy and restless, hardly staying on a tree for a few minutes; feeds on berries, fruits, but also hunts insects in a flycatcher manner. **FOOD** forest berries, fruit, insects, flower nectar. **VOICE** very noisy; an assortment of whistles and screeches. **DISTRIBUTION** several races; resident in Himalayas and NE India. Southern race *H.l. ganeesa* is darker with square tail and is found in W Ghats and Sri Lanka. **HABITAT** tall forests; hill-station gardens.

Plain Martin ■ *Riparia paludicola* 12cm

DESCRIPTION Sexes alike. Long wings and slight tail-fork. Grey-brown above, slightly darker on crown; dark brown wings and tail; dull grey below, whiter towards abdomen. The **Sand Martin** *R. riparia* (13cm) is white below, with a broad, grey-brown band across breast. A gregarious species, always in flocks, flying around sand-banks along water courses; individual birds occasionally stray far and high; hawks small insects in flight; flocks perch on telegraph wires. **FOOD** insects captured in night. **VOICE** a *brret...* call, rather harsh in tone, usually on the wing around nest colony; twittering song. **DISTRIBUTION** NW and N India, from outer Himalaya, south at least to line from vicinity of Mumbai-Nasik to C Orissa; moves considerably locally. **HABITAT** vicinity of water, sandbanks, sandy cliffsides.

Sand Martin

Dusky Crag Martin ■ *Ptyonoprogne concolor* 13cm

DESCRIPTION Sexes alike. Dark sooty-brown above; square-cut, short tail, with white spot on all but outermost and central tail feathers; paler underbody; faintly rufous chin and throat, with indistinct black streaking. The **Eurasian Crag Martin** *P. rupestris* is slightly larger and a paler sandy-brown, much paler below; Breeds in the NW and W Himalaya and winters in N and C India. Small parties; flies around ruins, crags and old buildings, hawking insects in flight; acrobatic, swallow-like flight and appearance; rests during hot hours on rocky ledges or some corner. **FOOD** insects, captured on wing. **VOICE** faint *chip ..*, uncommonly uttered. **DISTRIBUTION** nearly all of India, south of Himalayan foothills, to about 1,500m. **HABITAT** vicinity of old forts, ruins, old stony buildings in towns.

ABOVE: *Eurasian Crag Martin*

Red-rumped Swallow ■ *Cecropis daurica* 19cm

DESCRIPTION Sexes alike. Glossy steel-blue above; chestnut supercilium, sides of head, neck-collar and rump; dull rufous-white below, streaked brown; deeply forked tail

diagnostic. Small parties spend much of the day on the wing; the migrant, winter-visiting race *nipalensis*, is highly gregarious; hawks insects along with other birds; freely perches on overhead wires, thin branches of bushes and trees; hunts insects amongst the most crowded areas of towns, over markets and refuse heaps, flying with amazing agility, wheeling and banking and stooping with remarkable mastery. **FOOD** insects caught on the wing. **VOICE** mournful chirping note; pleasant twittering song of breeding male. **DISTRIBUTION** 6 races over the subcontinent, including Sri Lanka; resident and migratory. **HABITAT** cultivation, vicinity of human habitation, town centres, rocky hilly areas.

Wire-tailed Swallow ■ *Hirundo smithii* 14cm

DESCRIPTION Sexes alike. Glistening steel-blue above; chestnut cap; unmarked, pure white underbody distinctive; two long, wire-like projections (tail-wires) from outer tail

feathers diagnostic. Solitary or small parties; almost always seen around water, either perched on overhead wires or hawking insects in graceful, acrobatic flight, swooping and banking; often flies very low, drinking from the surface; roosts in reed beds and other vegetation, often with warblers and wagtails. **FOOD** insects captured on wing. **VOICE** soft twittering note; pleasant song of breeding male. **DISTRIBUTION** common breeding (summer) visitor to N India, to about 1,800m in the Himalaya; breeds in many other parts of India; widespread over the area, excepting arid zones. **HABITAT** open areas, cultivation, habitation, mostly in vicinity of canals, lakes, rivers.

Chestnut-headed Tesia ▪ *Oligura castaneocoronata* 8cm

DESCRIPTION Tiny bird with a dumpy body, long legs and a stubby tail; olive-green back and a chestnut hood; bright lemon-yellow throat and underparts olive-washed yellow; sides of breast and flanks olive-green. Shy and elusive, jerks body when calling; keeps to the ground, hopping around in bushy undergrowth. **FOOD** insects and spiders. **VOICE** chattering *chirruk chirruk…* or loud, piercing *tzit…*, repeated when alarmed. **DISTRIBUTION** resident in Himalayas and NE Indian hills, up to 3,900m. **HABITAT** thick undergrowth in moist forest; dark ravines near streams; moss-covered boulders or logs.

Grey-sided Bush Warbler ▪ *Cettia brunnifrons* 10cm

DESCRIPTION Olive-brown with bright rufous cap; long pale supercilium (well defined in front of the eye) with dark eye-stripe; whitish throat and belly with conspicuous grey breast and sides; vent olive-brown. Usually remains close to the ground, feeding in the undergrowth, tail sometimes cocked when disturbed. **FOOD** Insects, seeds. **VOICE** soft *tsik tsik*. **DISTRIBUTION** altitudinal migrant, breeding up to 4,000m in Himalayas, winters below 2,200m. **HABITAT** rhododendron, shrubberies and bushes at forest edges, open forest, tea gardens in winter.

Paddyfield Warbler ■ *Acrocephalus agricola* 13cm

DESCRIPTION Sexes alike. Rufescent-brown above; brighter on rump; whitish throat, rich buffy below. **Blyth's Reed Warbler** *A. dumetorum* (14cm) is a very common winter visitor; also has a whitish throat and buffy under body, but olive-brown upper body is distinctive. Solitary, hopping amidst low growth; rarely seen along with other birds; damp areas, especially reed growth and cultivation are favourite haunts; flies low, but soon vanishes into growth. **FOOD** insects. **VOICE** a *chrr…chuck* or a single *chack* note, rather harsh in tone. Blyth's Reed Warbler has a somewhat louder, quicker *tchik…* or *tchi…tchi…* call, and rarely, a warbling song before migration, around early April. **DISTRIBUTION** winter visitor; common over most of India, south of and including the terai. **HABITAT** damp areas, reed growth, tall cultivation.

ABOVE: *Blyth's Reed Warbler*

Clamorous Reed Warbler ■ *Acrocephalus stentoreus* 19cm

DESCRIPTION Sexes alike. Brown above; distinct pale supercilium; whitish throat, dull buffy-white below; at close range, or in the hand, salmon-coloured inside of mouth; calls diagnostic. Solitary or in pairs; difficult to see but easily heard; elusive bird, keeping to dense low reeds, mangrove and low growth, always in and around water; never associates with other species; flies low, immediately vanishes into the vegetation; occasionally emerges on reed or bush tops, warbling with throat puffed out. **FOOD** insects. **VOICE** highly vocal; loud *chack, chakrrr* and *khe* notes; distinctive, loud warbling; loud, lively song. **DISTRIBUTION** from Kashmir valley, south through the country; sporadically breeds in many areas, migrant in others. **HABITAT** reed beds, mangrove.

Booted Warbler
■ *Iduna caligata* 12cm

DESCRIPTION Sexes alike. Dull olive-brown above; short, pale white supercilium; pale buffy-white below. Blyth's Reed Warbler is brighter olive-brown and mostly frequents bushes. Solitary or two to four birds, sometimes in mixed bands of small birds; very active and agile, hunting amongst leaves and upper branches; overall behaviour very leaf-warbler like, but calls diagnostic. **FOOD** insects. **VOICE** harsh, but low *chak… chak…churr* calls almost throughout day; soft, jingling song, sometimes heard before departure in winter grounds. **DISTRIBUTION** winters over peninsula south from Punjab to West Bengal; breeds in NW regions, parts of W Punjab. **HABITAT** open country with *acacias* and scrub; occasionally light forests.

Green-crowned Warbler ■ *Seicercus burkii* 10cm

DESCRIPTION Sexes alike. Olive-green above; greenish or grey-green eyebrow bordered above with prominent black coronal bands; greenish sides of face, yellow eye-ring; completely yellow below. The **White-Spectacled Warbler** S. *affinis* has grey on crown and whitish eye-ring. Small, restless flocks, often in association with other small birds; keeps to the low bush and lower branches of trees. **FOOD** insects. **VOICE** fairly noisy; sharp *chip… chip…* or *cheup…cheup…* notes. **DISTRIBUTION** breeds in the Himalaya, 2,000–3,000m; winters in foothills, parts of C and E peninsula, N Maharashtra, S Madhya Pradesh, NE Andhra Pradesh. **HABITAT** forest undergrowth.

White-spectacled Warbler

Grey-hooded Warbler ■ *Seicercus xanthoschistos* 10cm

DESCRIPTION Sexes alike. Grey above; prominent, long white eyebrow; yellowish rump and wings; white in outer tail seen in flight; completely yellow below. The **Grey-cheeked Warbler** *S. poliogenys* has a dark slaty head, white eye-ring and grey chin and cheeks; it is found in Himalaya, east of Nepal. Pairs or small bands, often along with mixed hunting parties; actively hunts and flits in canopy foliage and tall bushes; highly energetic. **FOOD** insects; rarely small berries. **VOICE** quite vocal; familiar calls in Himalayan forests; loud, high-pitched, double-note call; pleasant, trilling song. **DISTRIBUTION** the Himalaya 900–3,000m; altitudinal movement in winter. **HABITAT** Himalayan forests, gardens.

ABOVE: *Grey-cheeked Warbler*

Large-billed Leaf Warbler ■ *Phylloscopus magnirostris* 12cm

DESCRIPTION Sexes alike. Brown-olive above; yellowish supercilium, dark eye-stripe distinctive; one or two faint wing-bars, not always easily seen; dull-yellow below. The very similar **Greenish Warbler** *P. trochiloides* (10cm) is best identifiable in field by its call (squeaky, fairly loud *dhciewee* or a *cheee…ee*). Usually solitary, sometimes in mixed parties of small birds; quite active; spends most time in leafy upper branches of medium-sized trees; not easy to sight, but characteristic call notes help in confirming its presence. **FOOD** small insects. **VOICE** distinctive *dir…tee…* call, the first note slightly lower; loud, ringing, 5-noted song. **DISTRIBUTION** breeds in the Himalaya, 1,800–3,600m; winters over most of the peninsula, though exact range imperfectly known. **HABITAT** forests, groves.

ABOVE: *Greenish Warbler*

Zitting Cisticola ■ *Cisticola juncidis* 10cm

DESCRIPTION Sexes alike. Rufous-brown above, prominently streaked darker; rufous-buff, unstreaked rump; white tips to fan-shaped tail diagnostic; buffy-white underbody, more rufous on flanks. Diagnostic calls. Pairs or several birds over open expanse; great skulker, lurking in low growth; usually seen during short, jerky flights, low over ground; soon dives into cover; most active when breeding, during rains; striking display of male, soaring erratically, falling and rising, incessantly uttering sharp, creaking note; adults arrive on nest in similar fashion. **FOOD** insects, spiders, possibly some seeds. **VOICE** sharp, clicking *zit...zit* calls; continuously during display in air. **DISTRIBUTION** subcontinent, south of the Himalayan foothills; absent in extreme NW Rajasthan. **HABITAT** open country, grass, cultivation, reed beds; also coastal lagoons.

Common Tailorbird ■ *Orthotomus sutorius* 13cm

DESCRIPTION Sexes alike. Olive-green above; rust-red fore-crown; buffy-white underbody; dark spot on throat sides, best seen in calling male; long, pointed tail, often held erect; central tail feathers about 5cm longer and pointed in breeding male. One of India's best-known birds; usually in pairs together; rather common amidst habitation, but keeps to bushes in gardens; remains unseen even when at arm's length, but very vocal; tail often cocked, carried almost to the back; clambers up into trees more than other related warblers. **FOOD** insects, flower nectar. **VOICE** very vocal; loud familiar *towit...towit*; song is a rapid version of call, with slight change, loud *chuvee...chuvee...chuvee*, uttered for up to 7 minutes at a stretch; male sings on exposed perch. **DISTRIBUTION** subcontinent, to about 2,000m in outer Himalaya. **HABITAT** forest, cultivation, habitation.

Grey-breasted Prinia ▪ *Prinia hodgsonii* 11cm

DESCRIPTION Sexes alike. Grey-brown, with some rufous above; long grey tail, tipped black and white; white underbody; when breeding, soft grey breast-band diagnostic. The **Rufous-fronted Prinia** *P. buchanani* (12cm) can be identified by the rufous head and dark brown tail, tipped black and white. Small bands ever on the move; keeps to low growth but often clambers into middle levels; singing males may climb to top of trees; few nearly always present in mixed hunting parties of small birds; nest like tailorbird's. **FOOD**

ABOVE: *Rufous-fronted Prinia*

insects, flower nectar. **VOICE** noisy when breeding; longish, squeaky song; contact calls, almost continuous squeaking. **DISTRIBUTION** all India south of Himalayan foothills up to about 1,800m; absent in arid W Rajasthan. **HABITAT** edges of forests, cultivation, gardens, scrub, often in and around habitation.

Graceful Prinia ▪ *Prinia gracilis* 13cm

DESCRIPTION Sexes alike. Dull grey-brown above, streaked darker; very pale around eyes; long, graduated tail, faintly cross-barred, tipped white; whitish underbody, buffy on belly. Plumage more rufous in winter. The **Striated Prinia** *P. criniger* (16cm) is larger, dark brown and streaked. **Rufous-vented Prinia** *P. burnesii* (17cm) is rufous-brown above, streaked; whitish below. Small parties move in low growth; usually does not associate with other birds; restless, flicks wings and tail often; occasionally hunts like flycatcher. **FOOD** insects. **VOICE** longish warble when breeding; wing-snapping and jumping display of male; *szeep...szip...* call note. **DISTRIBUTION** NW Himalayan foothills, terai, south to Gujarat, across Gangetic plain. **HABITAT** scrub, grass, canal banks, semi-desert.

Graceful Prinia   *Striated Prinia*   *Rufous-vented Prinia*

Plain Prinia ■ *Prinia inornata* 13cm

DESCRIPTION Sexes alike. Pale brown above; whitish
supercilium and lores; dark wings and tail; long, graduated
tail, with buff tips and white outer feathers; buff-white
underbody; tawny flanks and belly. In winter, more rufous
above. The **Yellow-bellied Prinia** *P. flaviventris* is olivish-
green above, with a slaty-grey head;
yellow belly and whitish throat
distinctive. Pairs or several move about
in low growth; skulker, difficult to see;
jerky, low flight, soon vanishing into
bush; tail often flicked. **FOOD** insects,
flower nectar. **VOICE** plaintive *tee…
tee*; also a *krrik…krrik* sound; wheezy
song, very insect-like in quality.
DISTRIBUTION subcontinent, from
terai and Gangetic plain southwards;
absent in W Rajasthan. **HABITAT**
tall cultivation, grass, scrub; prefers
damp areas.

Yellow-bellied Prinia

Ashy Prinia ■ *Prinia socialis* 13cm

DESCRIPTION Sexes alike. Rich, ashy-grey
above, with rufous wings and long, white-
tipped tail; whitish lores; dull buffy-rufous
below. In winter, less ashy, more rufous-
brown; longer tail; whitish chin and throat.
Mostly in pairs; common and familiar as
Common Tailorbird in some areas; actively
moves in undergrowth; often flicks and
erects tail; typical jerky flight when flying
from bush to bush; noisy and excited when
breeding. **FOOD** insects, flower nectar.
VOICE nasal *pee…pee…pee…*; song, a loud
and lively *jivee…jivee…jivee…* or *jimmy…
jimmy…*, rather like Common Tailorbird's
in quality, but easily identifiable once
heard. **DISTRIBUTION** subcontinent
south of Himalayan foothills, up to about
1,400m; absent in W Rajasthan. **HABITAT**
cultivation, edges of forest, scrub, parks,
vicinity of habitation.

Rusty-cheeked Scimitar Babbler ■ *Pomatorhinus erythrogenys* 25cm

DESCRIPTION Sexes alike. Olive-brown above; orangish-rufous (rusty) sides of face, head, thighs and flanks; remainder of underbody mostly pure white; long, curved 'scimitar' beak. Small bands in forest; a bird of undergrowth, hopping on jungle floor; turns over leaves or digs with beak; sometimes hops into leafy branches, but more at ease on ground. **FOOD** insects, grubs, seeds. **VOICE** noisy; mellow, fluty whistle, 2-noted *cue...pe...cue...pe*, followed by single (sometimes double) note reply by mate; guttural alarm call and a liquid contact note. **DISTRIBUTION** Himalaya foothills to at least 2,200m and possibly to 2,600m. **HABITAT** forest undergrowth, ravines, bamboo.

Indian Scimitar Babbler ■ *Pomatorhinus horsfieldii* 22cm

DESCRIPTION Sexes alike. Deep olive-brown above; long white supercilium; white throat, breast and belly-centre; long, curved yellow 'scimitar' beak. Pairs or small, loose bands in forest; keeps to undergrowth, where the bubbling, fluty calls are heard more often than the birds are seen; hops on jungle floor, vigorously rummaging amidst leaf litter, digs with long beak; hops its way into leafy branches, but not for long; scattered birds keep in touch through calls. The similar **White-browed Scimitar Babbler** *P.schisticeps* is found in the Himalaya. **FOOD** insects, spiders, flower nectar. **VOICE** fluty, musical whistle, often followed by a bubbling note; often calls in duet. **DISTRIBUTION** hilly forest regions of peninsular India, with four races. **HABITAT** mixed forest, scrub, bamboo.

ABOVE: *White-browed Scimitar Babbler*

Slender-billed Scimitar Babbler ■ *Xiphirhynchus superciliaris* 20cm

DESCRIPTION Sexes alike. Very long, slender, sharply down-curved black bill; relatively small, dark grey head with a long white supercilium; dark brown above and brighter rufous-brown below; whitish throat streaked with grey. Skulking and secretive; forages in bamboo or in low undergrowth; often in small flocks. **FOOD** insects, berries. **VOICE** varied calls include a series of repeated powerful mellow hoots uttered rapidly. **DISTRIBUTION** resident in E Himalayas and NE India. **HABITAT** moist broadleaved forest, bamboo thickets.

Tawny-bellied Babbler
■ *Dumetia hyperythra* 13cm

DESCRIPTION Sexes alike. Olivish-brown above; reddish-brown front part of crown; white throat in western and southern races; nominate race has underbody entirely fulvous. Small, noisy parties in undergrowth; rummages on floor, hopping about, always wary; hardly associates with other birds; great skulkers, difficult to see, any sign of danger and the flock disperses amidst a noisy chorus of alarm notes but soon reunite. **FOOD** chiefly insects, but occasionally seen on flowering silk-cotton trees; also other flower nectar. **VOICE** faint *cheep…cheep* contact notes; also a mix of other whistling and chattering notes. **DISTRIBUTION** from SE Himachal Pradesh, east along foothills into peninsular India; absent in arid NW, Punjab plains, extreme NE states. **HABITAT** scrub and bamboo, in and around forests.

Yellow-eyed Babbler ■ *Chrysomma sinense* 18cm

DESCRIPTION Sexes alike. Rufous-brown above; whitish lores, short supercilium; yellow eye (iris) and orange-yellow eye-rim distinctive at close range; cinnamon wings; long, graduated tail; white below, tinged pale fulvous on flanks and abdomen. Pairs or small bands in tall grass and undergrowth; noisy but skulking, suddenly clambering into view for a few seconds, before vanishing once again; works its way along stems and leaves, hunting insects; short, jerky flight. **FOOD** insects, larvae; also flower nectar. **VOICE** noisy when breeding (mostly rains); melodious, whistling notes; also a mournful *cheep...cheep* call. **DISTRIBUTION** subcontinent, from the Himalayan foothills south; absent in arid parts of Rajasthan. **HABITAT** scrub, tall grass, cultivation, edges of forest.

Puff-throated Babbler
■ *Pellorneum ruficeps* 15cm

DESCRIPTION Sexes alike. Olivish-brown above; dark rufous cap; whitish-buff stripe over eye; white throat; dull fulvous-white underbody, boldly streaked blackish-brown on breast and sides. Solitary or in pairs; shy, secretive bird of undergrowth; mostly heard, extremely difficult to see; rummages on ground, amidst leaf litter; hops about, rarely ascends into upper branches. **FOOD** insects. **VOICE** noisy when breeding; mellow whistle, 2, 3, or 4-noted; best-known call is a 4-note whistle, interpreted as *he-will-beat-you*. **DISTRIBUTION** hilly-forest areas; Himalaya, to about 1,500m, east of SE Himachal Pradesh; NE states; S Bihar, Orissa, Satpura range across C India, E and W Ghats. **HABITAT** forest undergrowth, bamboo, overgrown ravines, nullahs.

Common Babbler
■ *Turdoides caudata* 23cm

DESCRIPTION Sexes alike. Dull brown above, profusely streaked; brown wings; olivish-brown tail long and graduated, cross-rayed darker; dull white throat; pale fulvous underbody, streaked on breast sides. Pairs or small bands in open scrub; skulker, working its way low in bush or on ground; moves with peculiar bouncing hop on the ground, the long, loose-looking tail cocked up; extremely wary, vanishing into scrub at slightest alarm; weak flight, evident when flock moves from one scrub patch to another, in ones and twos. **FOOD** insects, flower nectar, berries. **VOICE** noisy; pleasant, warbling whistles, several birds often in chorus; squeaky alarm notes; calls on ground and in low flight. **DISTRIBUTION** most of N, NW, W and peninsular India, south of outer Himalaya to about 2,000m; east to about West Bengal. **HABITAT** thorn scrub, open cultivation, grass.

Striated Babbler ■ *Turdoides earlei* 21cm

DESCRIPTION Sexes alike. Dull brownish above, streaked darker; long, cross-barred tail; buffy-brown below, with fine dark streaks on throat and breast (the Common Babbler has a white throat and lacks breast streaks). The **Striated Grassbird** *Megalurus plaustris* (25cm), with greatly overlapping range, has bolder streaking above and a prominent whitish supercilium, and is almost white below, streaked below breast. Sociable; parties of up to 10 birds keep to tall grass and reed beds; flies low, rarely drops down to the ground. **FOOD** insects, snails. **VOICE** loud, 3-noted whistle, also a quick-repeated, single whistling note. **DISTRIBUTION** floodplains of N and NE river systems, especially the larger rivers. **HABITAT** tall grass, reed beds and scrub.

Striated Grassbird

Large Grey Babbler ■ *Turdoides malcolmi* 28cm

DESCRIPTION Sexes alike. Grey-brown above; dark centres to feathers on back give streaked look; greyer forehead; long graduated tail cross-rayed with white outer feathers, conspicuous in flight; fulvous-grey below. Gregarious; flocks in open country, sometimes dozens together; extremely noisy; moves on ground and in medium-sized trees; hops about, turning over leaves on ground; weak flight, never for long; at any sign of danger, the flock comes together. **FOOD** insects, seeds, berries; rarely flower nectar. **VOICE** very noisy; a chorus of squeaking chatter; short alarm note. **DISTRIBUTION** from around E Uttar Pradesh, Delhi environs, south through most of peninsula; east to Bihar; abundant in the Deccan. **HABITAT** scrub, open country, gardens, vicinity of habitation.

Yellow-billed Babbler ■ *Turdoides affinis* 24cm

DESCRIPTION Sexes alike. Creamy-white crown; dull brown above, appearing scaly on centre of back; darker wings and cross-barring along tail centre; dark brown throat and breast, the pale grey edges to feathers giving scaled appearance; yellowish-buff below breast. Small noisy parties;

feeds on ground, turning leaves; if disturbed, moves about in a series of short, hopping flights; hops amongst tree branches towards the top, from where a short flight takes the birds to an adjoining tree. **FOOD** insects, nectar, figs. **VOICE** noisy; definitely more musical chatter than the more common Jungle Babbler. **DISTRIBUTION** southern peninsular India-Karnataka, Andhra, Tamil Nadu, Kerala; also Sri Lanka. **HABITAT** forests, dense growth, neighbourhood of cultivation and habitation, orchards.

White-crested Laughingthrush ■ *Garrulax leucolophus* 28cm

DESCRIPTION Sexes alike. Olive-brown above; pure white head, crest, throat, breast and sides of head; broad black band through eye to ear-coverts; rich rufous nuchal collar, continuing around breast; olive-brown below breast. Small parties in forest; moves in undergrowth but readily ascends into upper leafy branches; makes short flights between

trees; very noisy; often seen along with other laughingthrushes, treepies and drongos; hops on ground, rummaging in leaf litter. **FOOD** insects, berries. **VOICE** very noisy; sudden explosive chatter or 'laughter'; also pleasant 2- or 3-note whistling calls. **DISTRIBUTION** the Himalaya, east of N Himachal Pradesh; foothills to a height of 2,400m, most common between 600–1,200m. **HABITAT** dense forest undergrowth, bamboo, wooded nullahs.

Black-faced Laughingthrush ■ *Garrulax affinis* 22cm

DESCRIPTION Sexes alike. Diagnostic blackish face, throat and part of head and contrasting white malar patches, neck sides and part of eye-ring; rufous-brown above,

finely scalloped on back; olivish-golden flight feathers tipped grey; rufous-brown below throat, marked grey. Pairs or small bands, sometimes with other babblers; moves on ground and in low growth; also ascends into middle levels of trees; noisy when disturbed or when snakes or other creatures arouse its curiosity. **FOOD** insects, berries, seeds. **VOICE** various high-pitched notes, chuckles; a rolling *whirrr* alarm call; a 4-noted, somewhat plaintive song. **DISTRIBUTION** the Himalaya, from W Nepal eastwards, descends to 1,500m in winter. **HABITAT** undergrowth in forest; dwarf vegetation in higher regions.

Variegated Laughingthrush ■ *Garrulax variegatus* 28cm

DESCRIPTION Sexes alike. Olive-brown above; grey, black and white head and face; grey, black, white and rufous in wings and tail; black chin and throat, bordered with buffy-white; narrow white tip to tail, with grey subterminal band. Small flocks, up to a dozen and more on steep, bushy hillsides; keeps to undergrowth for most part, but occasionally clambers into leafy branches; wary and secretive, not easily seen; weak flight, as in most laughingthrushes. **FOOD** insects, fruits; rarely flower nectar. **VOICE** noisy; clear, musical whistling notes, 3 to 4 syllables; also harsh, squeaking notes. **DISTRIBUTION** the Himalaya, east to C Nepal; 1,200–3,500m; breeds between 2,000–3,200m. **HABITAT** forests undergrowth, bamboo; seen in hill-station gardens in winter.

White-throated Laughingthrush ■ *Garrulax albogularis* 28cm

DESCRIPTION Sexes alike. Greyish olive-brown above, fulvous forehead, black mark in front of eye, full rounded tail with 4 outer pairs of feathers broadly tipped with white. Rufous below but with conspicuous pure white throat sharply demarcated by a line of olive-brown. The white gorget stands out in the gloom of forest floor. **FOOD** insects; also berries. **VOICE** continual chattering; warning *twit-tzee* alarm. **DISTRIBUTION** throughout Himalaya, with distinct western race *whistleri*, up to 3,000m in summer. **HABITAT** dense forest, scrub, wooded ravines.

Striated Laughingthrush
▪ *Garrulax striatus* 28cm

DESCRIPTION Sexes alike. Rich-brown plumage, heavily white-streaked, except on wings and rich rufous-brown tail; darkish, loose crest, streaked white towards front; heavy streaking on throat and sides of head, becoming less from breast downwards. Pairs or small parties; often along with other birds in mixed, noisy parties; feeds both in upper branches and in low bushes; shows marked preference for certain sites in forest. **FOOD** insects, fruits; seen eating leaves. **VOICE** very vocal; clear whistling call of 6- to 8-notes; loud, cackling chatter. **DISTRIBUTION** 800–2,700m; the Himalaya east of Kulu, parts of NE states. **HABITAT** dense forests, scrub, wooded ravines.

Streaked Laughingthrush
▪ *Garrulax lineatus* 20cm

DESCRIPTION Sexes alike. Pale grey plumage, streaked dark brown on upper back, white on lower back; rufous ear-coverts and wings; rufous edges and grey-white tips to roundish tail; rufous streaking and white shafts on underbody. Pairs or small bands; prefers low bush and grassy areas, only rarely going into upper branches; hops, dips and bows about; flicks wings and jerks tail often; weak, short flight. **FOOD** insects, berries; refuse around hillside habitation. **VOICE** fairly noisy; a near-constant chatter of a mix of whistling and squeaky notes; common call a whistle of 2 or 3 notes, *pitt...wee...er*. **DISTRIBUTION** the Himalaya, west to east; 1,400–3,800m; considerable altitudinal movement. **HABITAT** bushy hill-slopes, cultivation, edges of forest.

Red-billed Leiothrix ▪ *Leiothrix lutea* 13cm

DESCRIPTION Male: olive-grey above; dull buffy-yellow lores and eye-ring; yellow, orange, crimson and black in wings; forked tail, with black tip and edges; yellow throat, orange-yellow breast diagnostic; scarlet beak. The red on the wing is considerably reduced or absent in the western race *kumaiensis*. Female: like male, but yellow instead of crimson in wings. Small parties, often a part of mixed hunting parties of small birds in forest;

rummages in undergrowth but frequently moves up into leafy branches; a lively, noisy bird. **FOOD** insects, berries. **VOICE** quite vocal; often utters a wistful, piping *tee... tee...tee*; a mix of sudden explosive notes; song a musical warble. **DISTRIBUTION** the Himalaya, from Kashmir to extreme NE; 600–2,700m. **HABITAT** forest undergrowth, bushy hillsides, plantations.

Rufous Sibia ▪ *Malacias capistratus* 20cm

DESCRIPTION Sexes alike. Rich-rufous plumage; grey-brown centre of back (between wings); black crown; slightly bushy crest and sides of head; bluish-grey wings and black shoulder-patch; grey-tipped long tail; black subterminal tail-band. Small flocks, sometimes

with other birds; active gymnasts, ever on the move; cheerful calls; hunts in canopy and middle forest levels, moves amidst moss-covered branches; springs into air after winged insects; sometimes hunts like treecreepers on stems, probing bark crevices. **FOOD** insects, flower nectar, berries. **VOICE** wide range of whistling and sharp notes; rich song of six to eight syllables during Himalayan summer. **DISTRIBUTION** Himalaya, 1,500–3,000m. Sometimes up to about 3,500m; descends to 600m in some winters. **HABITAT** forest, both temperate and broadleafed.

White-browed Fulvetta
■ *Fulvetta vinipectus* 11cm

DESCRIPTION Sexes alike. Brown crown and nape; prominent white eyebrow with black or dark brown line above; blackish sides of face; olive-brown above, washed rufous on wings, rump and tail; some grey in wings; whitish throat and breast; olive-brown below. Up to 20 birds in low growth or lower branches; energetic, acrobatic birds, often seen in mixed hunting parties. **FOOD** insects, caterpillars, berries. **VOICE** fairly sharp *tsuip...* or *tship...* call; also some harsh churring notes, when agitated. **DISTRIBUTION** the Himalaya from W Himachal Pradesh; E Himalaya; NE regions; 1,500–3,500m, over 4,000m in some parts; descends to 1,200m in severe winters. **HABITAT** scrub in forest, *ringal* bamboo.

Brown-cheeked Fulvetta ■ *Alcippe poioicephala* 15cm

DESCRIPTION Sexes alike. Olive-brown above; grey crown and neck; thin black stripe through eye; rufescent-brown wings and tail; dull fulvous underbody. Pairs or small parties,

often along with other birds; moves actively in undergrowth and leafy branches, clinging sideways or springing from perch; rather shy in most areas, but occasionally emerges into open areas. **FOOD** insects, spiders, flower nectar, berries. **VOICE** best-known call is the 4- to 8-syllabled song, interpreted as *daddy-give-me-chocolate*; harsh *churrr...* notes serve as contact calls. **DISTRIBUTION** peninsular India, south from S Rajasthan across Pachmarhi (Satpuras) to S Bihar and Orissa. **HABITAT** forests, undergrowth, bamboo; also hill-station gardens in W Ghats.

Whiskered Yuhina ■ *Yuhina flavicollis* 13cm

DESCRIPTION Sexes alike. Olive-brown above; chocolate-brown crown and crest; white eye-ring and black moustache seen from close up; rufous-yellow nuchal collar (less distinct in western race *albicollis*); white underbody, streaked rufous-olive on sides of breast and flanks. Flocks, almost always in association with other small birds; active and restless, flitting about or hunting flycatcher-style; moves between undergrowth and middle levels of forests; sometimes ascending into canopy; keeps up a constant twitter. **FOOD** insects, berries, flower nectar. **VOICE** quite vocal; a mix of soft twittering notes and fairly loud titmice-like 2- or 3-note call, *chee…chi…chew*. **DISTRIBUTION** the Himalaya; W Himachal Pradesh to extreme NE; 800–3,400m. **HABITAT** forests.

Oriental White-eye ■ *Zosterops palpebrosus* 10cm

DESCRIPTION Sexes alike. Olive-yellow above; short blackish stripe through eye; white eye-ring distinctive; bright yellow throat and under tail; whitish breast and belly. Small parties, occasionally up to 40 birds, either by themselves or in association with other small birds; keeps to foliage and bushes; actively moves amongst leafy branches, clinging sideways

and upside-down; checks through leaves and sprigs for insects and also spends considerable time at flowers; calls often, both when in branches and when flying in small bands from tree to tree. **FOOD** insects, flower nectar, berries. **VOICE** soft, plaintive *tsee…* and *tseer…* notes; short jingling song. **DISTRIBUTION** all India, from Himalayan foothills to 2,000m; absent in arid parts of W Rajasthan. **HABITAT** forests, gardens, groves, secondary growth.

Asian Fairy Bluebird ■ *Irena puella* 28cm

DESCRIPTION Male: glistening blue above; deep velvet-black sides of face, underbody and wings; blue under tail-coverts. Female: verditer-blue plumage; dull black lores and flight feathers. Pairs or small loose bands; spends the day in leafy tall branches; often descends into undergrowth to feed on berries or hunt insects; utters their 2-noted calls while flitting amongst trees; seen along with other birds. **FOOD** fruits, insects, flower nectar. **VOICE** common call a double-noted *wit...weet...*; also a *whi...chu...*; besides, some harsh notes occasionally heard. **DISTRIBUTION** disjunct distribution; one population in W Ghats and associated hills south of Ratnagiri; another in E Himalaya, east of extreme SE Nepal; Uttarakhand foothills; Andaman and Nicobar Islands. **HABITAT** dense, evergreen forests, sholas.

Chestnut-bellied Nuthatch ■ *Sitta (castanea) cinnamoventris* 12cm

DESCRIPTION Male: blue-grey above; black stripe from lores to nape; whitish cheeks and upper throat; all but central tail feathers black, with white markings; chestnut below. Female: duller chestnut below. The male **White-tailed Nuthatch** *S. himalayensis* has a much paler underbody and clear white patch at base of tail. Pairs or several, often with other small birds; restless climber; clings to bark and usually works up the tree stem, hammering with beak; also moves upside-down and sideways; may visit the ground. **FOOD** insects, grubs, seeds. **VOICE** loud *tzsib...* call; faint twitter, loud whistle during breeding season. **DISTRIBUTION** lower Himalaya east of Uttarakhand; to about 1,800m; the recently split **Indian Nuthatch** *S. castanea* is found in the peninsula. **HABITAT** forests, groves, roadside trees, habitation.

White-tailed Nuthatch

Velvet-fronted Nuthatch
▪ *Sitta frontalis* 10cm

DESCRIPTION Male: violet-blue above; jet-black forehead; stripe through eye; white chin and throat, merging into vinous-grey below; coral-red beak. Female: lacks black stripe through eye. Pairs or several in mixed hunting parties; creeps about on stems and branches; fond of moss-covered trees; also clings upside-down; active and agile, quickly moving from tree to tree; calls often, till long after sunset; also checks fallen logs and felled branches. **FOOD** insects. **VOICE** fairly loud, rapidly repeated, sharp, trilling *chweet…chwit…chwit* whistles. **DISTRIBUTION** from around W Uttarakhand east along lower Himalaya; widespread over the hilly, forested areas of C, S and E India; absent in the flat and arid regions. **HABITAT** forests; also tea and coffee plantations.

Bar-tailed Treecreeper
▪ *Certhia himalayana* 12cm

DESCRIPTION Sexes alike. Streaked blackish-brown, fulvous and grey above; pale supercilium; broad fulvous wing-band; white chin and throat; dull ash-brown below; best recognized by dark brown barring on pointed tail. Solitary or several in mixed parties of small birds; spends almost entire life on tree-trunks; starts climbing from near the base; intermittently checks crevices and under moss, picks out insects with curved beak; usually climbs to mid-height, then moves on to another tree; sometimes creeps on moss-clothed rocks and walls. **FOOD** insects, spiders. **VOICE** long-drawn squeak, somewhat ventriloquial; loud but short, monotonous song; one of the earliest bird songs, heard much before other birds have begun to sing. **DISTRIBUTION** the Himalaya, east to W Nepal; from about 1,600m to timberline; descends in winter. **HABITAT** Himalayan temperate forests.

Common Hill Myna
▪ *Gracula religiosa* 28cm

DESCRIPTION Sexes alike. Black plumage, with a purple-green gloss; white in flight feathers; orange-red beak; orange-yellow legs, facial skin and fleshy wattles on nape and sides of face. In the **Lesser Hill Myna** G. *indica*, nape-wattles extend up along sides of crown, the eye and nape wattles distinctly separated. Small flocks in forest; extremely noisy; mostly arboreal, only occasionally descending into bush or onto ground; hops amongst branches, and on ground; large numbers gather on fruiting trees, along with barbets, hornbills and green pigeons. Such sights are one of the birdwatching spectacles of the Himalayan foothills. **FOOD** fruits, insects, flower nectar, lizards. **VOICE** amazing vocalist; great assortment of whistling, warbling, shrieking notes; excellent mimic; much sought-after cage bird. **DISTRIBUTION** is found along lower Himalaya and terai, from Uttarakhand eastwards; the Lesser Hill Myna is a bird of the W Ghats, from North Kanara to extreme south, and Sri Lanka; the race *peninsularis* is restricted to Orissa, E Madhya Pradesh and adjoining N Andhra Pradesh. **HABITAT** forests, clearings.

Lesser Hill Myna

Common Myna ▪ *Acridotheres tristis* 23cm

DESCRIPTION Sexes alike. Rich vinous-brown plumage; black head, neck and upper breast; yellow beak, legs and naked wattle around eyes distinctive; large white spot in dark brown flight feathers, best seen in flight; blackish tail, with broad white tips to all but central feathers; whitish abdomen. Solitary, or in scattered pairs or small, loose bands India's most common and familiar bird; hardly ever strays far from man and habitation; rather haughty and confident in looks; aggressive, curious and noisy; struts about on ground, picks out worms; attends to grazing cattle and refuse dumps; enters verandahs and kitchens, sometimes even helping itself on dining tables. **FOOD** omnivorous; fruits, nectar, insects, kitchen scraps, refuse. **VOICE** noisy; a great mix of chattering notes, one of India's most familiar bird sounds. **DISTRIBUTION** subcontinent, up to about 3,500m in the Himalaya. **HABITAT** human habitation, cultivation, light forests.

Bank Myna

▪ *Acridotheres ginginianus* 23cm

DESCRIPTION Sexes alike. Similar to Common Myna but smaller; has bluish-grey neck, mantle and underparts; black head with orange-red wattle around the eye; orange-yellow bill; buff-orange tail-tips and wing-patch. Usually observed in small, scattered groups around human habitation; bold and confiding; often seen along roadside restaurants picking out scraps. **FOOD** omnivorous; fruits, nectar, insects, kitchen scraps, refuse. **VOICE** similar to Common Myna's but softer. **DISTRIBUTION** widespread resident in N and C India. **HABITAT** human habitation, cultivation, grassland.

Common Starling ▪ *Sturnus vulgaris* 20cm

DESCRIPTION Glossy black plumage, with iridescent purple and green; plumage spotted with buff and white; hackled feathers on head, neck and breast; yellowish beak and red-brown legs. Summer (breeding) plumage mostly blackish. Several races winter in N India, with head purple or bronze-green, but field identification of races not very easy in winter.

Gregarious, restless birds; feeds on ground, moving hurriedly, digging with beak in soil; entire flock may often take off from ground; flies around erratically or circles, but soon settles on trees or returns to ground. **FOOD** insects, berries, grain, earthworms, small lizards. **VOICE** mix of squeaking, clicking notes; other chuckling calls. **DISTRIBUTION** the race *indicus* breeds in Kashmir to about 2,000m; this and three other races winter over NW and N India, occasionally straying south to Gujarat; quite common in parts of N India in winter. **HABITAT** meadows, orchards, vicinity of habitation, open, fallow land.

Asian Pied Starling ■ *Gracupica contra* 23cm

DESCRIPTION Sexes alike. Black and white (pied) plumage distinctive; orange-red beak and orbital skin in front of eyes confirm identity. Sociable; small parties either move on their own or associate with other birds, notably other mynas and drongos; rather common and familiar over its range but keeps a distance from man; may make its ungainly nest

in garden trees, but never inside houses, nor does it enter houses; more a bird of open, cultivated areas, preferably where there is water; attends to grazing cattle; occasionally raids standing crops. **FOOD** insects, flower nectar, grain. **VOICE** noisy; a mix of pleasant whistling and screaming notes. **DISTRIBUTION** bird of north-central, central and eastern India, south and east of a line roughly from E Punjab, through E Rajasthan, W Madhya Pradesh to the Krishna delta; escaped cage birds have established themselves in several areas out of original range as in and around Mumbai. **HABITAT** open cultivation, orchards, vicinity of habitation.

Chestnut-tailed Starling

■ *Sturnia malabarica* 21cm

DESCRIPTION Sexes alike. Silvery-grey above, with faint brownish wash; dull rufous till breast, brighter below; black and grey in wings. Sociable; noisy parties in upper branches of trees, frequently along with other birds; incessantly squabbles and moves about, indulging in all manners of acrobatic positions to obtain nectar or reach out to fruit; descends to ground to pick up insects. **FOOD** flower nectar, fruits, insects. **VOICE** noisy, metallic, whistling call, becoming a chatter when there is a flock; warbling song when breeding.

DISTRIBUTION India, roughly east and south from S Rajasthan to around W Uttarakhand; up to about 1,800m in Himalayan foothills. **Blyth's Starling** *S. m. blythii* breeds in SW India, Karnataka and Kerala, spreading north to Maharastra in winter. **HABITAT** light forest, open country, gardens.

Blyth's Starling

Brahminy Starling
■ *Sturnia pagodarum* 20cm

DESCRIPTION Sexes alike. A grey, black and rufous myna; black crown, head and crest; grey back; rich-buff sides of head, neck and underbody; black wings and brown tail with white sides and tip distinctive in flight. Female has a slightly smaller crest, otherwise like male. Small parties, occasionally collecting into flocks of 20 birds; associates with other birds on flowering trees or on open lands; walks typical myna style, head held straight up, confident in looks; communal roosting-sites, with other birds. **FOOD** fruits, flower nectar, insects. **VOICE** quite noisy; pleasant mix of chirping notes and whistles, sounding as conversational chatter; good mimic; pleasant warbling song of breeding males. **DISTRIBUTION** subcontinent, to about 2,000m in W and C Himalaya. **HABITAT** light forests, gardens, cultivation, vicinity of habitation.

Rosy Starling ■ *Pastor roseus* 24cm

DESCRIPTION Sexes alike. Rose-pink and black plumage; glossy black head, crest, neck, throat, upper breast, wings and tail; rest of plumage rose-pink, brighter with the approach of spring migration. Gregarious; flocks often contain young birds, crestless, dull brown

and sooty; often along with other mynas on flowering *Erythrina* and *Bombax* trees; causes enormous damage to standing crops; seen also around grazing cattle and damp open lands; overall an aggressive and extremely noisy bird; huge roosting colonies, resulting in deafening clamour before settling. **FOOD** grain, insects, flower nectar. **VOICE** very noisy; mix of guttural screams, chattering sounds and melodious whistles. **DISTRIBUTION** winter visitor to India, particularly common in N, W and C India; arrives as early as end of July; most birds depart around mid-April to early May; absent or uncommon east of Bihar. **HABITAT** open areas, cultivation, orchards, flowering trees amidst habitation.

Malabar Whistling Thrush ■ *Myophonus horsfieldii* 25cm

DESCRIPTION Sexes alike. Deep blue-black plumage, more glistening on wings and tail; bright, cobalt-blue forehead and shoulder-patch. Solitary or in pairs; a lively bird of hilly, forested country; keeps to forest streams and waterfalls; also perches on trees; peculiar stretching of legs and raising of tail; often encountered on roadside culverts, from where it bolts into nullah or valley. **FOOD** insects, crustaceans, snails, frogs, berries. **VOICE** renowned vocalist; especially vocal during the rains; begins to call very early in morning; a rich, whistling song, very human in quality, nicknamed 'whistling schoolboy'; its fluty notes float over the roar of water; also, a harsh, high-pitched *kreeee* call. **DISTRIBUTION** hills of W India, from S Rajasthan, south all along W Ghats, to about 2,200m; also parts of Satpuras. **HABITAT** forest streams, waterfalls, gardens.

Orange-headed Thrush
■ *Zoothera citrina* 21cm

DESCRIPTION Blue-grey above; orangish-rufous head, nape and underbody; white ear-coverts with two dark brown vertical stripes; white throat and shoulder patch. The Orange-headed nominate race has entire head rufous-orange. Usually in pairs; feeds on ground, rummaging in leaf litter and under thick growth; flies into leafy branch if disturbed; occasionally associates with laughingthrushes and babblers; vocal and restless when breeding. **FOOD** insects, slugs, small fruit. **VOICE** loud, rich song, often with a mix of other birds' calls thrown in; noisy in early mornings and late evenings, also a shrill, screechy *kreeee...* call. **DISTRIBUTION** peninsular India south of a line from S Gujarat across to Orissa; the nominate race breeds in the Himalaya, NE; winters in foothills, terai, parts of E India, Gangetic plains and south along E Ghats. **HABITAT** shaded forests, bamboo groves, gardens.

Tickell's Thrush ▪ *Turdus unicolor* 22cm

DESCRIPTION Male: light ashy-grey plumage; duller breast and whiter on belly; rufous underwing-coverts in flight. Female: olive-brown above; white throat, streaked on sides; tawny flanks and white belly. Small flocks on the ground, sometimes along with other thrushes; hops fast on ground, stopping abruptly, as if to check some underground activity; digs worms from under soil; flies into trees when approached too close. **FOOD** insects, worms, small fruit. **VOICE** rich song; double-noted alarm call; also some chattering calls. **DISTRIBUTION** breeds in the Himalaya, 1,500–2,500m, east to C Nepal, and Sikkim; winters along foothills east of Kangra, NE, and parts of C and E peninsular India. **HABITAT** open forests, groves.

Indian Blackbird ▪ *Turdus (merula) simillimus* 25cm

DESCRIPTION Male: lead-grey above, more ashy-brown below; blackish cap distinctive; darker wings and tail; reddish-orange beak and yellow eye-rims distinctive. Female: dark ashy-brown above; browner below, with a grey wash; streaked dark brown on chin and throat. The black-capped race *T. m. nigropileus* of the W Ghats, S Rajasthan and parts of Vindhyas has a more distinct black cap; in the **Tibetan Blackbird** *T. m. maximus* (27cm), the male is entirely black with a yellow beak; female is dark brown. Solitary or pairs, sometimes with other birds; rummages on forest floor but also moves up in leafy branches; rather confiding, especially in hill-station gardens. **FOOD** insects, small fruit, earthworms. **VOICE** loud, melodious song of breeding male; sings from high tree perch; very vocal in evening; great mimic; screechy *kreeee* during winter; also a harsh *charr* note. **DISTRIBUTION** various races make this a widespread species in the Indian region; hills of W India, from S Rajasthan southwards; E Ghats south of N Orissa. **HABITAT** forests, ravines, gardens.

Long-tailed Thrush ■ *Zoothera dixoni* 27cm

DESCRIPTION Plain olive-brown above; two dull buffy wing-bars and a larger wing-patch seen best in flight; buffy throat, breast and flanks, the rest white, boldly spotted dark brown. The confusingly similar **Plain-backed Thrush** Z. *mollissima* has very indistinct wing-bars; it has a shorter tail but this character is not very useful in the field. The **Scaly Thrush** Z. *dauma* (26cm) has a distinctly spotted back. Pairs or several together in winter; feeds on ground; usually difficult to spot till it takes off from somewhere close by; flies up into branches if disturbed. **FOOD** insects, snails. **VOICE** mostly silent; *mollissima* has a loud, rattling alarm note.

DISTRIBUTION the Himalaya, east of C Himachal Pradesh; breeds between 2,000–4,000m; descends to about 1,000m in winter. **HABITAT** timberline forest; scrub in summer; heavy forests in winter

Plain-backed Thrush

White-tailed Rubythroat ■ *Luscinia pectoralis* 15cm

DESCRIPTION Male: slaty above; white supercilium; white in tail; scarlet chin and throat; jet-black sides of throat, continuing into broad breast-band, white below, greyer on sides. Female: grey-brown above; white chin and throat; greyish breast. The **Siberian Rubythroat** L. *calliope* male lacks black on breast; has white malar stripe; female has a brown breast; winters in NE and E India. Solitary; wary; difficult to observe; cocks tail; hops on ground, or makes short dashes; ascends small bush tops. **FOOD** insects, molluscs. **VOICE** short, metallic call note; short, harsh alarm note; rich, shrill song.

DISTRIBUTION breeds in the Himalaya, 2,700–4,600m; winters in N, NE India; winter range not properly known. **HABITAT** dwarf vegetation, rocky hills in summer. In winter, prefers cultivation, damp ground with grass and bush.

Siberian Rubythroat

Indian Blue Robin ▪ *Luscinia brunnea* 15cm

DESCRIPTION Male: deep slaty-blue above; white supercilium; blackish lores and cheeks; rich-chestnut throat, breast and flanks; white belly centre and under tail. Female: brown above; white throat and belly; buffy-rufous breast and flanks. The **White-browed Bush Robin** *Tarsiger indicus* male has a very long, conspicuous supercilium, and completely rufous-orange underbody; it is resident in the Himalaya. Solitary, rarely in pairs; great skulker, very difficult to observe; moves amidst dense growth and hops on ground; jerks and flicks tail and wings often. **FOOD** insects. **VOICE** high-pitched *churr* and harsh *tack…* in winter; trilling song of breeding male, sometimes singing from exposed perch. **DISTRIBUTION** breeds in the Himalaya, 1,500–3,300m. Winters in southern W Ghats, Ashambu Hills, and Sri Lanka. **HABITAT** dense rhododendron, *ringal* bamboo undergrowth in summer. Evergreen forest undergrowth, coffee estates in winter.

ABOVE: *White-browed Bush Robin*

Oriental Magpie Robin ▪ *Copsychus saularis* 20cm

DESCRIPTION Male: glossy blue-black and white; white wing-patch and white in outer tail distinctive; glossy blue-black throat and breast; white below. Female: rich slaty grey, where male is black. A familiar bird of India. Solitary or in pairs, sometimes with other birds in mixed parties; hops on ground, preferring shaded areas; common about habitation; when perched, often cocks tail; flicks tail often, especially when making short sallies; active at dusk; remarkable songster, very rich voice. **FOOD** insects, berries, flower nectar. **VOICE** one of India's finest songsters; rich, clear song of varying notes and tones; male sings from exposed perches, most frequently between March and June, intermittently year round; also has harsh *churr* and *chhekh* notes; a plaintive *sweee…* is a common call. **DISTRIBUTION** subcontinent, up to about 1,500m in outer Himalaya; absent in extreme W Rajasthan. **HABITAT** forests, parks, towns.

White-rumped Shama ■ *Copsychus malabaricus* 25cm

DESCRIPTION Male: glossy-black head and back; white rump and sides of graduated tail distinctive; black throat and breast; orange-rufous below. Female: grey where male is black; slightly shorter tail and duller rufous below breast. Usually pairs; overall behaviour like Oriental Magpie Robin's; arboreal bird of forest, hill-station gardens; keeps to shaded areas and foliage, only occasionally emerging in open; launches short sallies and hunts till late in evening. **FOOD** insects; rarely flower nectar. **VOICE** rich songster; melodious, three or four whistling notes very characteristic; variety of call notes, including a mix of some harsh notes. **DISTRIBUTION** Himalayan foothills, terai, east of Uttarakhand; NE India; hill forests of Bihar, Orissa, SE Madhya Pradesh, E Maharashtra, south along E Ghats to about Cauvery river; entire W Ghats, from Kerala north to S Gujarat. **HABITAT** forests, bamboo, hill station gardens.

Indian Robin ■ *Saxicoloides fulicatus* 16cm

DESCRIPTION Several races in India. Males differ in having dark brown, blackish-brown or glossy blue-back upper body. Male: dark brown above; white wing-patch; glossy blue-black below; chestnut vent and under tail. Female: lacks white in wings; duller grey-brown below. Solitary or in pairs in open country, and often in and around habitation; rather suspicious and maintains safe distance between man and itself; hunts on ground, hopping or running in short spurts; when on ground, holds head high and often cocks tail, right up to back, flashing the chestnut vent and under tail. **FOOD** insects. **VOICE** long-drawn *sweeeech* or *weeeech* call; a warbling song when breeding; also a guttural *charrr…* note. **DISTRIBUTION** subcontinent, south of the Himalayan foothills; absent in extreme NE. **HABITAT** open country, edges of forest, vicinity of habitation, scrub.

Black Redstart
■ *Phoenicurus ochruros* 15cm

DESCRIPTION Male: black above (marked with grey in winter); grey crown and lower back; rufous rump and sides of tail; black throat and breast; rufous below. Female: dull brown above; tail as in male; dull tawny-brown below. The eastern race *rufiventris* has a black crown, and is the common wintering bird of India. Mostly solitary in winter, when common all over India; easy bird to observe, in winter and in its open high-altitude summer country; perches on overhead wires, poles, rocks and stumps; characteristic shivering of tail and jerky body movements; makes short dashes to ground, soon returning to perch with catch; rather confiding in summer, breeding in houses, under roofs and in wall crevices. **FOOD** insects, mostly taken on ground. **VOICE** squeaking *tictititic...* call, often beginning with faint *tsip...* note; trilling song of breeding male. **DISTRIBUTION** breeds in the Himalaya, 2,400–5,200m; winters over much of subcontinent. **HABITAT** open country, cultivation.

Blue-fronted Redstart ■ *Phoenicurus frontalis* 15cm

DESCRIPTION Male: bright blue forehead with darker blue crown and back; orange-chestnut underparts; rufous rump; orange tail with broad blackish terminal band and central feathers. Female: dark olive-brown; yellowish-orange below; rump and tail as in male; tail pattern diagnostic, to separate it from other female redstarts. Mostly solitary, perched on rocks or bushes; drops

to the ground to feed; pumps tail. **FOOD** insects, seeds, berries on the ground. **VOICE** squeaking *tik* or *prik*. **DISTRIBUTION** altitudinal migrant; breeds in Himalayas up to 5,300m; winters in the Himalayan foothills. **HABITAT** cultivation, open country, gardens.

Plumbeous Water Redstart ■ *Rhyacornis fuliginosa* 12cm

DESCRIPTION Male: slaty-blue plumage; chestnut tail diagnostic; rufous on lower belly. Female: darkish blue-grey-brown above; two spotted wing-bars; white in tail; whitish below, profusely mottled slaty. Young birds are brown, also with white in tail. Pairs on mountain rivers; active birds, making short dashes from boulders; move from boulder to boulder, flying low over roaring waters; tail frequently fanned open and wagged; hunts late in evening; maintains feeding territories in winter as well as at other times. **FOOD** insects, worms. **VOICE** sharp *kree*... call; also a snapping *tzit...tzit*; rich, jingling song of breeding male, infrequently uttered in winter. **DISTRIBUTION** the Himalaya, 800–4,000m, but mostly 1,000–2,800m; also breeds south of Brahmaputra river; in winter may descend into the foothills, terai. **HABITAT** mountain streams, rivers, rushing torrents.

White-capped Water Redstart ■ *Chaimarrornis leucocephalus* 19cm

DESCRIPTION Sexes alike. Black back, sides of head, wings and breast; white crown diagnostic; chestnut rump and tail; black terminal tail-band; chestnut below breast. The male **Guldenstadt's Redstart** *Phoenicurus erythrogaster* (16cm) has a completely chestnut tail and prominent white wing-patch. Solitary or pairs on Himalayan torrents; rests on rocks amidst gushing waters, flying very low over the waters to catch insects; jerks and wags tail and dips body; restless bird; interesting display of courting male. **FOOD** insects. **VOICE** loud, plaintive *tseeee* call; also a *psit...psit...* call; whistling song of breeding male. **DISTRIBUTION** Himalaya: 2,000–5,000m; descends into foothills in winter. **HABITAT** rocky streams; also on canals in winter.

LEFT: *Guldenstadt's Redstart*

Little Forktail ■ *Enicurus scouleri* 12cm

DESCRIPTION Sexes alike. Black and white plumage. Black above, with white forehead; white band in wings extends across lower back; small, black rump-patch; slightly

forked, short tail with white in outer feathers; black throat, white below. Solitary or in pairs; a bird of mountain streams, waterfalls and small, shaded forest puddles; energetically moves on moss-covered and wet, slippery rocks; constantly wags and flicks tail; occasionally launches short sallies, but also plunges underwater, dipper style. **FOOD** aquatic insects. **VOICE** rather silent save for a rarely uttered sharp *tzittzit* call. **DISTRIBUTION** the Himalaya, west to east. Breeds between 1,200–3,700m; descends to about 300m in winter. **HABITAT** rocky mountain streams, waterfalls.

Spotted Forktail ■ *Enicurus maculatus* 25cm

DESCRIPTION Sexes alike. White forehead and fore-crown; black crown and nape; black back spotted white; broad white wing bar and rump; deeply forked, graduated black

and white tail; black till breast, white below. The white-spotted back easily distinguishes this species from other similar sized forktails in the Himalaya. Solitary or in scattered pairs; active bird, moving on mossy boulders at water's edge or in mid-stream; long, forked tail gracefully swayed, almost always kept horizontal; flies low over streams, calling; sometimes rests in shade of forest; commonly seen bird of the Himalaya. **FOOD** aquatic insects, molluscs. **VOICE** shrill, screechy *kree* call, mostly in flight; also some shrill, squeaky notes on perch. **DISTRIBUTION** the Himalaya; breeds mostly between 1,200–3,600m; descends to about 600m in winter. **HABITAT** boulder-strewn torrents, forest streams, roadsides.

Common Stonechat
■ *Saxicola torquatus* 13cm

DESCRIPTION Male: black above; white
rump, wing-patch and sides of neck/breast
(collar); black throat; pinkish-orange breast.
In winter, black feathers broadly edged
buff-rufous-brown. Female: rufous-brown
above, streaked darker; unmarked yellowish-
brown below; white wing-patch and rufous
rump. Solitary or in pairs in open country;
perches on small bush tops, fence-posts
and boulders; restless, makes short trips to
ground to capture insects, soon returning to
perch. FOOD insects. VOICE double-noted
wheet chat call; soft, trilling song of breeding
male in Himalaya, occasionally in winter
grounds. DISTRIBUTION breeds in Himalaya,
1,500–3,000m; winters all India except Kerala
and much of Tamil Nadu. HABITAT dry, open
areas, cultivation, tidal creeks.

Pied Bushchat ■ *Saxicola caprata* 13cm

DESCRIPTION Male: black plumage;
white in wing, rump and belly. Female:
brown above, paler on lores; darker tail;
dull yellow-brown below, with a rusty
wash on breast and belly. Solitary or
in pairs; perches on a bush, overhead
wire, pole or some earth mound; makes
short sallies on to ground, either
devouring prey on ground or carrying
it to perch; active, sometimes guards
feeding territories in winter; flicks and
spreads wings; fascinating display flight
of courting male (April–May). FOOD
insects. VOICE harsh, double-noted
call serves as contact and alarm call;
short, trilling song of breeding male.
DISTRIBUTION subcontinent, from
outer Himalaya to about 1,500m.
HABITAT open country, scrub,
cultivation, ravines.

Grey Bushchat ■ *Saxicola ferrea* 15cm

DESCRIPTION Male: dark grey above, streaked black; black mask; white supercilium, wing-patch and outer tail; white throat and belly; dull grey breast. Female: rufous-brown, streaked; rusty rump and outer tail; white throat; yellow-brown below. Solitary or pairs; like other chats, keeps to open country and edge of forest; perches on bush tops and poles, flirts tail often; regularly seen in an area; flies to ground on spotting insect. **FOOD** insects. **VOICE** double-noted call; also a grating *praee...* call; trilling song of male. **DISTRIBUTION** Himalaya, 1,400–3,500m; descends into foothills and adjoining plains, including Gangetic plains, in winter. **HABITAT** open scrub, forest edges, cultivation.

Desert Wheatear ■ *Oenanthe deserti* 15cm

DESCRIPTION Male: sandy above, with whitish rump and black tail; black wings; white in coverts; black throat and head-sides; creamy-white below. Female: brown wings and tail; lacks black throat. Winter male: throat feathers fringed white. The **Isabelline Wheatear** *O. isabellina* (16cm) is larger and sandy-grey, without black throat. The male **Northern Wheatear** *O. oenanthe* is grey above, with a white rump and tail sides and black tail centre and tip like an inverted 'T'; black ear-coverts and wings. Keeps to ground or perches on low bush or small rock; has favoured haunts; colouration makes it difficult to spot; makes short sallies to capture insects. **FOOD** insects. **VOICE** in winter an occasional *ch...chett* alarm note; reportedly utters its short, plaintive song in winter too. **DISTRIBUTION** winter visitor over N, C and W India, almost absent south of S Maharashtra and Andhra Pradesh; the Tibetan race *oreophila* breeds in Kashmir, Ladakh, Lahaul and Spiti, at about 3,000–5,000m. **HABITAT** open rocky, barren country; sandy areas; fallow lands.

Isabelline Wheatear

Northern Wheatear

Brown Rock Chat
▪ *Cercomela fusca* 17cm

DESCRIPTION Brown above, more rufous below; dark brown wings, almost blackish tail. Overall appearance like female Indian Robin. Usually pairs, around ruins, dusty villages, rocky hillsides; often approaches close; tame and confiding; captures insects on ground; rather aggressive when breeding. **FOOD** insects; occasionally kitchen refuse. **VOICE** harsh *chaeck...* note; also a whistling *chee* call; melodious song of breeding male; a good mimic. **DISTRIBUTION** confined to parts of N and C India, from Punjab and W Uttarakhand, south to about Narmada river; east to the Bihar-Bengal. **HABITAT** dry, open country, rocky hills, ravines, ruins, habitation.

Blue Rock Thrush ▪ *Monticola solitarius* 23cm

DESCRIPTION Male: blue plumage; brown wings and tail; pale fulvous and black scales more conspicuous in winter; belly whiter in winter. Female: duller, grey-brown above; dark shaft-streaks; black barring on rump; dull white below, barred brown. Solitary; has favoured sites, often around habitation; perches on rocks, stumps, roof tops; has a rather upright posture; flies on to ground to feed, but sometimes launches short aerial sallies.

FOOD insects, berries; rarely flower nectar. **VOICE** silent in winter; short, whistling song of breeding male. **DISTRIBUTION** breeds in the Himalaya, from extreme west to end Nepal; 1,200–3,000m, perhaps higher; winters from foothills, NE, south throughout peninsula; uncommon in Gangetic plains. **HABITAT** open rocky country, cliffs, ravines, ruins, habitation.

Chestnut-bellied Rock Thrush ■ *Monticola rufiventris* 23cm

DESCRIPTION Male: cobalt-blue head and upperparts with blackish mask; rich chestnut belly. Female: olive-brown with buff throat and lores; heavy scaling on underparts; distinctive face pattern with eye-ring, dark malar stripe and neck-patch. Mostly solitary or seen in pairs; perches upright. **FOOD** insects, berries. **VOICE** harsh rattle, fluty song. **DISTRIBUTION** Himalayan forests up to 3,500m. **HABITAT** open country, forest edges, groves on rocky hillsides.

Blue-capped Rock Thrush

■ *Monticola cinclorhynchus* 17cm

DESCRIPTION Male: blue crown and nape; black back; broad stripe through eyes to ear-coverts; blue throat and shoulder-patch; white wing-patch and chestnut rump distinctive; chestnut below throat. Back feathers edged fulvous in winter. Female: unmarked olive-brown above; buffy-white below, thickly speckled with dark brown. The female Blue Rock Thrush M. *solitarius* is grey-brown above, has a yellow-brown vent and a dull wing bar. Solitary or in pairs; an elusive forest bird; moves in foliage in mixed parties or rummages on ground, amidst leaf litter; best seen when it emerges in clearing. **FOOD** insects, flower nectar, berries. **VOICE** mostly silent in winter, save for an occasional harsh single or double-noted call; rich song of breeding male. **DISTRIBUTION** breeds in the Himalaya, 1,000–2,500m, sometimes higher; winters in W Ghats, from Narmada river south; sporadic winter records from C Indian forests. **HABITAT** shaded forests, groves.

Asian Brown Flycatcher ■ *Muscicapa dauurica* 13cm

DESCRIPTION Sexes alike; ashy-brown; greyish wash on dirty white breast; short tail; large head with a huge eye and prominent eye-ring; basal half of lower mandible pale and fleshy; black legs. **Brown-breasted Flycatcher** M. *muttui* is similar but has pronounced brown breast-band, and larger bill with entirely pale lower mandible. Usually solitary; perches upright on lower branches of trees, making sallies to catch insects and returning to the same perch. **FOOD** insects. **VOICE** call thin *tzee*, whistling song. **DISTRIBUTION** breeding resident of Himalayan foothills, with several small disjunct populations resident in the hills of central and peninsular India; widespread winter visitor in peninsula. **HABITAT** open forest, groves, gardens, plantations.

Brown-breasted Flycatcher

Rufous-gorgeted Flycatcher
■ *Ficedula strophiata* 14cm

DESCRIPTION Male: dark olive-brown upperparts; blackish face and throat; conspicuous white forehead and eyebrow; diagnostic rufous-orange gorget which is not always visible; grey breast; white sides to black tail. Female similar but duller, less distinct eyebrow and gorget. Frequently seen perched quietly in shaded areas or dense canopy. Like all flycatchers, hawks insects but sometimes feeds on the ground. **FOOD** insects. **VOICE** metallic pink, harsh *trrt*. **DISTRIBUTION** uncommon resident in Himalayas and NE Indian hills. **HABITAT** forest clearings and edges.

Ultramarine Flycatcher ■ *Ficedula superciliaris* 10cm

DESCRIPTION Male: deep blue above and sides of head, neck and breast, forming a broken breast-band; long white eyebrow; white in tail; white below. Female: dull-slaty above; grey-white below. The eastern race *aestigma* lacks white over eye and in tail. Solitary or in pairs; seen in mixed parties during winter; active, hunts in characteristic flycatcher style; rarely ventures into open. **FOOD** insects. **VOICE** faint *tick…tick…* in winter; a *chrrr* alarm note; three-syllabled song in the Himalaya.

DISTRIBUTION breeds in Himalaya, 1,800–3,200m; winters in N and C India, south to Karnataka and N Andhra Pradesh. **HABITAT** forests, groves, orchards, gardens.

Slaty-blue Flycatcher ■ *Ficedula tricolor* 13cm

DESCRIPTION Male: slaty-blue above; greyish-white (Himalayas) or buff (S Assam hills) below; black mask; white patch at base of black tail. Female: brown upperparts; warm brownish-buff flanks; rufescent rump and tail. Slim, long-tailed flycatcher; usually solitary or in pairs; feeds near the ground with tail cocked. **FOOD** insects. **VOICE** faint *tick tick* call. **DISTRIBUTION** breeds in Himalayas between 1,800 and 2,600m; winters in foothills. **HABITAT** forest undergrowth, reeds, bushes, grass.

Black-and-orange Flycatcher
■ *Ficedula nigrorufa* 13cm

DESCRIPTION Male: rich orangish-rufous plumage; blackish crown, nape, sides of face and wings. Female: like male, but deep olive-brown head; pale eye-ring. Usually solitary, but pairs often close by; not often seen in mixed hunting parties; keeps to dense, shaded undergrowth, either hopping low or making short flycatcher-like sallies from low perch; in its restricted range, quite tame and confiding once spotted. **FOOD** insects. **VOICE** soft, gloomy *pee…* call note; a sharp *zit…zit* alarm call; high-pitched, metallic song when breeding. **DISTRIBUTION** very local; restricted to Nilgiris and associated hills in southern W Ghats, most common above 1,500m. **HABITAT** dense evergreen forest, undergrowth, bamboo.

Verditer Flycatcher ■ *Eumyias thalassinus* 15cm

DESCRIPTION Male: verditer-blue plumage, darker in wings and tail; black lores. Female: duller, more grey overall. The **Nilgiri Flycatcher** *E. albicaudatus* of W Ghats is darker blue with white in tail; the **Pale Blue Flycatcher** *Cyornis unicolor* (16cm) male is uniform blue, with white on belly; female is olive-brown. Solitary or in pairs in winter, sometimes with other birds; restless, flicking tail; swoops about, ever on the move, occasionally descending quite low; rather more noticeable than other flycatchers because of its continuous movement and habit of perching in open exposed positions, like a bare twig on a tree top. **FOOD** insects. **VOICE** silent in winter, save for a rare, faint *chwe…* call; rich, trilling notes and song during Himalayan summer. **DISTRIBUTION** breeds in the Himalaya; 1,200-3,200m; winters in Indian plains, hill forests of C, E and S India. **HABITAT** open forests, orchards.

Nilgiri Flycatcher   *Pale Blue Flycatcher*

White-bellied Blue Flycatcher ■ *Cyornis pallipes* 15cm

DESCRIPTION Male: indigo-blue above; black lores; bright blue forehead, supercilium; indigo-blue throat and breast; white below breast. Female: deep olive-brown above; chestnut tail; rufous-orange till breast, whiter below. The female **Blue-throated Blue Flycatcher** C. *rubeculoides* (14cm) has a dull-rufescent (not chestnut) tail. Solitary, rarely in pairs; sometimes in mixed parties; mostly silent and unobtrusive, hence overlooked; hunts in low growth, often flicking tail. **FOOD** insects. **VOICE** soft 2-noted call; longish, squeaky song when breeding; rather silent for most part of the year. **DISTRIBUTION** W Ghats, south of C Maharashtra, around the latitude of Pune. **HABITAT** dense forest undergrowth.

Tickell's Blue Flycatcher
■ *Cyornis tickelliae* 14cm

DESCRIPTION Male: dark indigo-blue above; bright blue on forehead and supercilium; darker, almost appearing black, on sides of face; rufous-orange throat and breast; whitish below. Female: duller overall. Usually in pairs in shaded areas, often in mixed hunting parties; vicinity of wooded streams are favoured haunts; flits about intermittently or launches short sorties; has favourite perches; often breaks into fluty song. **FOOD** insects. **VOICE** clear, metallic song of 6 notes, sometimes extending to 9 or 10; often uttered in winter too. **DISTRIBUTION** all India, south roughly of a line from Kutch to W Uttarakhand east along terai; absent in extreme N, NW India. **HABITAT** shaded forests, bamboo, gardens.

Rufous-bellied Niltava
▪ *Niltava sundara* 15cm

DESCRIPTION Blue patch on sides of neck.
Male: deep purple-blue back and throat; dark
blue mask; black forehead; brilliant blue crown,
shoulders and rump; chestnut-rufous underbody.
Female: olivish-brown overall; rufescent tail;
white on lower throat diagnostic. The **Large
Niltava** *N. grandis* (21cm) male is dark blue
with tufted forehead. The **Blue-throated Blue
Flycatcher** male has dark blue throat and
white belly; it is duller, uniform blue above.
Mostly solitary; keeps to undergrowth; highly
unobtrusive, seldom seen; often flicks wings like
redstart, and bobs body. **FOOD** insects. **VOICE**
squeaky churring note; occasionally a sharp
psi…psi; also some harsh notes and squeaks.
DISTRIBUTION the Himalaya, NE; 1,500–
3,200m; winters in foothills, adjoining plains.
HABITAT dense forest undergrowth, bushes.

Blue-throated Blue Flycatcher   Large Niltava

Golden-fronted Leafbird ▪ *Chloropsis aurifrons* 19cm

DESCRIPTION Leaf-green plumage, golden-orange forehead; blue shoulder-patches; dark
blue chin (blackish in southern races;) and cheeks, black lores and ear-coverts, continuing
as a loop around blue throat. Pairs
in leafy canopy; lively bird, actively
hunting in foliage; its wide range of
whistling and harsh notes immediately
attracts attention; owing to greenish
plumage, difficult bird to see in foliage;
rather aggressive, driving away other
birds especially on flowering trees.
FOOD insects, spiders, flower nectar.
VOICE noisy; wide assortment of
whistling notes, including imitations
of several species; most common call
a drongo or shikra-like *che…chwe.*
DISTRIBUTION up to about 1,600m
in Uttarakhand Himalaya; east to
Bihar, Orissa, south along E Ghats and
up the W Ghats and adjoining areas.
HABITAT forests.

Thick-billed Flowerpecker
■ *Dicaeum agile* 9cm

DESCRIPTION Sexes alike. Olive-grey above, greener on rump; white-tipped tail; dull whitish-grey below, streaked brown, more on breast; orange-red eyes and thick, blue-grey beak seen at close range. Solitary or in pairs in canopy foliage; arboreal, restless; flicks tail often as it hunts under leaves or along branches; frequents parasitic clumps of *Loranthus* and *Viscum*. **FOOD** figs, berries of *Ficus, Lantana, Loranthus* and *Viscum*; also feeds on insects, spiders and nectar. **VOICE** loud, sharp *chik...chik*. **DISTRIBUTION** India south of and including the Himalayan foothills; absent over arid parts of NW India and from large tracts of Tamil Nadu. **HABITAT** forests, orchards, gardens.

Pale-billed Flowerpecker ■ *Dicaeum erythrorhynchos* 8cm

DESCRIPTION Sexes alike. Olive-brown above; unmarked grey-white below; pinkish-flesh and yellow-brown beak seen only at close range or in good light. The **Nilgiri Flowerpecker** *D. concolor* has a dark beak and a pale supercilium. Solitary or two to three birds in canopy; frequents parasitic *Loranthus* and *Viscum*; flits from clump to clump; strictly arboreal, restless; territorial even when feeding. **FOOD** causes much damage to orchards, especially mango and guava; chiefly berries, spiders and small insects. **VOICE** sharp, loud *chik...chik*; **DISTRIBUTION** from Kangra east along foothills to NE India; peninsular India south of a line from W Gujarat to S Bihar. **HABITAT** light forests, groves.

ABOVE: *Nilgiri Flowerpecker*

Fire-breasted Flowerpecker ■ *Dicaeum ignipectus* 7cm

DESCRIPTION Male: metallic blue-green-black above; buffy below, with scarlet breast-patch and black stripe down centre of lower breast and belly. Female: olive-green above, yellowish on rump; bright buff below; flanks and sides tinged olive. Mostly solitary; arboreal and active; flits about in foliage canopy, attending to *Loranthus* clumps; may be encountered in the restless mixed hunting bands of small birds in Himalayan forests. **FOOD** berries, nectar, spiders, small insects. **VOICE** sharp, metallic *chip...chip* note; high-pitched, clicking song. **DISTRIBUTION** the Himalaya, Kashmir to extreme east; breeds at 1,400–3,000m; winters as low as 300m. **HABITAT** forests, orchards.

Purple-rumped Sunbird ■ *Leptocoma zeylonica* 10cm

DESCRIPTION Male: deep chestnut-crimson back; metallic-green crown and shoulder-patch; metallic-purple rump and throat; maroon collar below throat; yellow below. Female: ashy-brown above, with rufous in wings; whitish throat; yellow below. Usually pairs; very active, flits from flower to flower; occasionally descends into flowering garden bushes. **FOOD** flower nectar, spiders, small insects. **VOICE** *tsiswee...tsiswee...* calls; sharp, twittering song of breeding male, much lower in tone and volume than that of Purple Sunbird. **DISTRIBUTION** peninsular India south of a line from around Mumbai, C Madhya Pradesh, S Bihar and West Bengal. **HABITAT** open forests, gardens, orchards; common in towns.

Purple Sunbird ▪ *Cinnyris asiaticus* 10cm

DESCRIPTION Br Male: metallic purple-blue above, and on throat and breast; dark purplish-black belly; narrow chestnut-maroon band between breast and belly; yellow and

scarlet pectoral tufts, normally hidden under wings. Female: olive-brown above; pale yellow below (*zeylonica* female has whitish throat). Non-breeding male: much like female but with a broad purple-black stripe down centre of throat to belly. Solitary or in pairs; an important pollinating agent, almost always seen around flowering trees and bushes; displays amazing agility and acrobatic prowess when feeding; sometimes hunts flycatcher style. **FOOD** nectar, small insects, spiders. **VOICE** more noisy than other sunbirds; loud *chweet...* notes. **DISTRIBUTION** subcontinent, south from Himalayan foothills to about 1,500m. **HABITAT** open forests, gardens, groves.

Mrs Gould's Sunbird
▪ *Aethopyga gouldiae* 10cm

DESCRIPTION Male: Strikingly coloured sunbird with rich red mantle and back; bright yellow underparts; purplish-blue crown and throat; metallic-blue tail; yellow rump. Female: olive-brown with yellow belly, vent and rump-band; grey crown and throat. Mostly solitary; sometimes in groups with Green-tailed Sunbird. **FOOD** nectar. **VOICE** sharp *tzit-tzit.* **DISTRIBUTION** resident in Himalayas and NE Indian hills. **HABITAT** rhododendrons, forest, gardens, scrub.

Green-tailed Sunbird ■ *Aethopyga nipalensis* 11cm

DESCRIPTION Male: dark metallic blue-green head and nape, bordered by maroon mantle; olive-green back and wings; metallic blue-green tail (appears dark); underparts bright yellow with red-streaked breast; yellow rump not always visible. NW Himalayan race *horsfieldi* has less maroon on mantle. Female: olive-green with greyish-olive throat; yellowish-olive on belly; rump slightly yellower than upperparts; pale tips to tail. **FOOD** nectar. **VOICE** Sharp *zig-zig*. **DISTRIBUTION** resident in Himalayas. **HABITAT** oak and rhododendron forests, scrub, gardens.

Crimson Sunbird ■ *Aethopyga siparaja* 15cm

DESCRIPTION Male: has long tail; metallic-green crown and tail; deep crimson back and neck sides; yellow rump not commonly seen; bright scarlet chin and breast; olive-yellow belly. Female: olive plumage, yellower below. In the W Ghats **Vigor's Sunbird** *A. s. vigorsii*, the male's breast is streaked yellow. Solitary or in pairs; active gymnast, hanging upside-down and sideways as it probes flowers; also hovers; moves a lot in forest, between tall bushes and canopy. **FOOD** nectar, small insects, spiders. **VOICE** sharp, clicking call notes; pleasant chirping song of breeding male (June–August). **DISTRIBUTION** Resident in Himalayas and hills of NE India; winters in Himalayan plains. **HABITAT** forests, gardens.

ABOVE: *Vigor's Sunbird*

Little Spiderhunter
■ *Arachnothera longirostra* 14cm

DESCRIPTION Sexes alike. Olive-green above; dark tail, tipped white; grey-white throat, merging into yellow-white below; orangish pectoral tufts. Very long, curved beak diagnostic. The much larger **Streaked Spiderhunter** A. *magna* (17cm) is olive-yellow, profusely streaked. Usually solitary; sometimes two or three birds in vicinity; active, moving considerably between bush and canopy; wild banana blossoms are a favourite, the bird clinging upside-down on the bracts; long, curved beak specially adapted to nectar diet. **FOOD** nectar; also insects and spiders. **VOICE** high-pitched *chee...chee* call; loud *which... which...* song, sounding somewhat like tailorbird song. **DISTRIBUTION** disjunct: found in both W Ghats and E Ghats, foothills from SE Nepal eastwards, E Himalaya and much of NE states. **HABITAT** forests, secondary growth, nullahs, sholas.

Streaked Spiderhunter

House Sparrow ■ *Passer domesticus* 15cm

DESCRIPTION Male: grey crown and rump; chestnut sides of neck and nape; black streaks on chestnut-rufous back; black chin, centre of throat and breast; white ear-coverts. The **Spanish Sparrow** P. *hispaniolensis* male has a chestnut crown and black streaks on flanks. Female: dull grey-brown above, streaked darker; dull whitish-brown below. Small parties to large gatherings; mostly commensal on man, feeding and nesting in and around habitation, including most crowded localities; also feeds in cultivation; hundreds roost together.

FOOD seeds; also insects, and often omnivorous. **VOICE** noisy; a medley of chirping notes; richer notes of breeding male; double- and triple-brooded. **DISTRIBUTION** subcontinent, to about 4,000m in the Himalaya. **HABITAT** habitation, cultivation.

ABOVE: *Spanish Sparrow*

Russet Sparrow ▪ *Passer rutilans* 15cm

DESCRIPTION Male: rufous-chestnut above, streaked black on back; whitish wing-bars; black chin and centre of throat, bordered with dull yellow. Female: brown above, streaked darker; pale supercilium and wings-bars; dull ashy-yellow below. The **Eurasian Tree Sparrow** P. *montanus* male has a black patch on white ear-coverts and lacks yellow on throat sides. Gregarious mountain bird; mostly feeds on ground, picking seeds; may associate with other finches; often perches on dry branches and overhead wires. **FOOD** seeds, insects. **VOICE** chirping notes *swee…* Indian Robin-like call. **DISTRIBUTION** the Himalaya; NE; breeds 1,200–2,600m, higher to about 4,000m in NE; descends in winter. **HABITAT** cultivation, edges of forest, mountain habitation.

Eurasian Tree Sparrow

Tibetan Snowfinch
▪ *Montifringilla adamsi* 17cm

DESCRIPTION Dull grey-brown above with some streaking on the back (less pronounced in juveniles and fresh plumage); blackish wing with white panel in wing-coverts; male has greyish-black bib (not usually seen in female); breeding male has black bill; in flight shows obvious white wing patch; white tail with central black feathers and narrow black terminal band. Gregarious and often seen in flocks or around human habitation. **FOOD** insects, seeds. **VOICE** calls include a hard *pink pink* and soft mewing. **DISTRIBUTION** breeding resident in N Himalayas at higher altitudes above 3,600m (lower in winters). **HABITAT** high-altitude scrub, meadows, rocky bushy slopes, hillsides.

Streaked Weaver ■ *Ploceus manyar* 15cm

DESCRIPTION Br Male: yellow crown; blackish head sides; fulvous streaks on dark brown back; heavily streaked lower throat and breast. Female and non-breeding male:

streaked above; yellow stripe over eye continues to behind ear-coverts; very pale below, boldly streaked on throat and breast. The **Black-breasted Weaver** *P. benghalensis* male has a dark breast-band. Gregarious; prefers tall grass and reed beds in well-watered areas; active, as a rule not flying into trees; often nests close to other weavers. **FOOD** seeds, grain, insects. **VOICE**

high-pitched chirping, wheezy notes and chatter, much like Baya Weaver's. **DISTRIBUTION** most of India south of the Himalaya; absent in parts of Rajasthan and NW regions; the eastern race *peguensis* is darker and much more rufous above. **HABITAT** reed beds, tall grass in well-watered areas, marshes.

ABOVE: *Black-breasted Weaver*

Baya Weaver ■ *Ploceus philippinus* 15cm

DESCRIPTION Breeding male: bright yellow crown; dark brown above, streaked yellow; dark brown ear-coverts and throat; yellow breast. Female: buffy-yellow above, streaked darker; pale supercilium and throat, turning buffy-yellow on breast, streaked on sides. Non-breeding male: bolder streaking than female; male of eastern race *burmanicus* has yellow

restricted to crown. Gregarious; one of the most familiar and common birds of India, best known for its nest; keeps to cultivated areas, interspersed with trees; feeds on ground and in standing crops. **FOOD** grain, seeds, insects, nectar. **VOICE** chirping and high-pitched wheezy notes of breeding male; very noisy at nest colony (monsoons). **DISTRIBUTION** most of India up to about 1,000m in outer Himalaya; absent in Kashmir. **HABITAT** open country, tree- and palm-dotted cultivation.

Red Avadavat
▪ *Amandava amandava* 10cm

DESCRIPTION Breeding male: crimson and brown, spotted white on wings and flanks; white-tipped tail. Female: brown above, spotted on wings; crimson rump; dull white throat; buffy-grey breast, yellow brown below. Non-breeding male: like female, but greyer throat; upper breast distinctive. Small flocks, often with other weavers; partial to tall grass and scrub, preferably around well-watered areas; active and vibrant birds and rather confiding; huge numbers captured for bird markets. **FOOD** grass seeds; also insects when breeding. **VOICE** shrill and high-pitched notes, also uttered in flight. **DISTRIBUTION** subcontinent, south of Himalayan foothills. **HABITAT** tall grass, reeds, sugar cane, scrub, gardens.

Indian Silverbill
▪ *Euodice malabarica* 10cm

DESCRIPTION Sexes alike. Dull-brown above, with white rump; very dark, almost black wings; pointed tail; pale buffy-white below, with some brown on flanks; thick, grey-blue or slaty beak striking. Gregarious; mostly keeps to scrub in open country; feeds on ground and on standing crops, especially millet; overall a rather 'dull' bird, both in colour and demeanour. **FOOD** small seeds, millet. **VOICE** faint *tee...tee* notes; sometimes also a whistling note. **DISTRIBUTION** subcontinent to about 1,500m in Himalaya, chiefly the outer ranges. **HABITAT** prefers dry areas; cultivation, scrub and grass; sometimes light, open forests.

Scaly-breasted Munia ■ *Lonchura punctulata* 10cm

DESCRIPTION Sexes alike. Chocolate-brown above; olivish-yellow, pointed tail; white bars on rump; chestnut sides of face, chin and throat; white below, thickly speckled with very dark brown on breast, flanks and part of belly (speckles may be absent during winter and much of summer). Sociable, moving in flocks of six to several dozen birds, often with other munias and weaver birds; feeds on ground and low bushes, but rests in trees. **FOOD** seeds, small berries; also insects. **VOICE** common call a double-noted *ki...tee...ki...tee*. **DISTRIBUTION** most of India, to about 1,500m in parts of the Himalaya; absent in much of Punjab, NW regions and W Rajasthan. **HABITAT** open scrub, cultivation, especially where interspersed with trees; also gardens.

Black-headed Munia ■ *Lonchura malacca* 10cm

DESCRIPTION Sexes alike. Black head, throat, breast, belly centre and thighs; rufous-chestnut back, deeper chestnut on rump; white upper belly and sides of underbody. The

Chestnut Munia *L. m. atricapilla* of N and NE India has white of lower parts replaced by chestnut. Gregarious, except when breeding, as in other munias; prefers reed beds and cultivation, especially where flooded; during breeding season (rains), often seen along with Streaked Weaver; feeds on ground. **FOOD** grass seeds,

ABOVE: *Chestnut Munia*

paddy; occasionally insects. **VOICE** faint *pee...pee...* calls. **DISTRIBUTION** foothills and terai from SE Punjab eastwards; most of NE, N Orissa; peninsular India south of line from Mumbai to S Madhya Pradesh. **HABITAT** reed beds, paddy, grass and scrub.

Robin Accentor ■ *Prunella rubeculoides* 17cm

DESCRIPTION Sexes alike. Pale brown above, streaked darker on back; grey head and throat; two whitish wing-bars; rufous breast and creamy-white belly; streaks on flanks. Flocks in winter occasionally along with other accentors, pipits and sparrows; rather tame and confiding around high-altitude habitation; hops on ground; flies into bushes if intruded upon beyond a point. **FOOD** insects, small seeds. **VOICE** a sharp trilling note; also a *tszi...tszi...*; short, chirping song. **DISTRIBUTION** high Himalaya; breeds 3,200–5,300m; descends in winter to about 2,000m, rarely below 1,500m. **HABITAT** Tibetan facies; damp grass, scrub; high-altitude habitation.

Rufous-breasted Accentor
■ *Prunella strophiata* 17cm

DESCRIPTION Sexes alike. Small chestnut and brown accentor; heavily streaked throat; brown-streaked crown and upperparts; orange supercilium with white in front of eyes and buffy-white moustache. **FOOD** insects, small seeds. **VOICE** sharp trilling note; also a *twitt...twitt...*; short, chirping song. **DISTRIBUTION** Himalayas: breeds between 2,700–5,000m; descends to about 1,200m in summer, rarely below 600m. **HABITAT** montane scrub, high-altitude habitation; descends lower in winter to bushy and fallow fields, nests in low bushes on hillsides.

Citrine Wagtail
■ *Motacilla citreola* 17cm

DESCRIPTION Grey back; diagnostic yellow head, sides of face, complete underbody; white in dark wings. The race *calcarata* has a deep-black back and rump; yellow of head may be paler in female; plumage of races often confusing. Sociable, often with other wagtails; shows marked preference for damp areas; sometimes moves on floating vegetation on pond surfaces; either walks cautiously or makes short dashes. **FOOD** insects, small snails. **VOICE** ordinary call note is a wheezy *tzzeep*, quite similar in tone to **Yellow Wagtail** M. *flava*. **DISTRIBUTION** winter visitor over most of India; the race *calcarata* breeds in Ladakh, Lahaul and Spiti and Kashmir, at 1,500–4,600m. **HABITAT** marshes, wet cultivation, jheel edges.

ABOVE: *Yellow Wagtail*

Grey Wagtail ■ *Motacilla cinerea* 17cm

DESCRIPTION Br Male: grey above; white supercilium; brownish wings, with yellow-white band; yellow-green at base of tail (rump); blackish tail with white outer feathers; black throat and white malar stripe; yellow below. Wintering male and female: whitish throat (sometimes mottled black in breeding female); paler yellow below. Mostly solitary or in pairs; typical wagtail, feeding on ground, incessantly wagging tail; settles on house roofs and overhead wires. **FOOD** insects, small molluscs. **VOICE** sharp *tzitsi…* calls, uttered on the wing; pleasant song and display flight of breeding male. **DISTRIBUTION** breeds in Himalaya, from N Baluchistan east to Nepal, 1,200–4,300m; winters from foothills south throughout India. **HABITAT** rocky mountain streams in summer; open areas, forest clearings, watersides in winter.

White-browed Wagtail
■ *Motacilla maderaspatensis* 21cm

DESCRIPTION Black above; prominent white supercilium and large wing-band; black throat and breast; white below. Female is usually browner where male is black. The black-backed races of **White Wagtail** M. *alba* have conspicuous white forehead. Mostly in pairs, though small parties may feed together in winter; a bird of flowing waters, being especially fond of rock-strewn rivers, though it may be seen on ponds and tanks; feeds at edge of water, wagging tail frequently; also rides on the ferry-boats plying rivers. **FOOD** insects. **VOICE** sharp *tzizit* or *cheezit…* call; pleasant whistling song of breeding male. **DISTRIBUTION** most of India south of Himalayan foothills to about 1,200m; only resident wagtail in the Indian plains, breeding up to 2,000m in peninsula mountains. **HABITAT** rocky streams, rivers, ponds, tanks; sometimes may enter wet cultivation.

White Wagtail

Paddyfield Pipit ■ *Anthus rufulus* 15cm

DESCRIPTION Sexes alike. Fulvous-brown above, with dark brown centres of feathers, giving a distinctive appearance; dark brown tail, with white outer feathers, easily seen in flight; dull-fulvous below, streaked dark brown on sides of throat, neck and entire breast. The winter-visiting Tawny Pipit A. *campestris* usually lacks streaks on underbody while Blyth's Pipit A. *godlewskii* is indistinguishable in field, except by its harsher call note. Pairs or several scattered on ground; run in short spurts; when disturbed, utters feeble note as it takes off; singing males perch on grass tufts and small bushes. **FOOD** insects, seeds, spiders. **VOICE** thin *tsip*, *tseep* and *tsip…tseep…* calls; trilling song of breeding male. **DISTRIBUTION** up to about 2,000m in outer Himalaya, south throughout India. **HABITAT** grassland, marshy ground, cultivation

ABOVE: *Tawny Pipit*

Blyth's Pipit

Olive-backed Pipit ▪ *Anthus hodgsoni* 15cm

DESCRIPTION Sexes alike. Olive-brown above, streaked dark brown; dull-white supercilium, two wing-bars and in outer-tail feathers; pale buff-white below, profusely streaked dark brown on entire breast and flanks. The **Tree Pipit** *A. trivialis* is brown above, without olive wash. Gregarious in winter; spends most time on ground, running briskly; if approached close, flies with *tseep…* call into trees; descends in a few minutes. **FOOD** insects, grass and other seeds. **VOICE** faint *tseep…* call; lark-like song of breeding male. **DISTRIBUTION** breeds in the Himalaya, east from W Himachal Pradesh; above 2,700m, to timberline; winters in foothills and almost all over India, except arid NW, Kutch; *trivialis* breeds only in NW Himalaya and is most common in winter over C India, but sporadically over range of *hodgsoni*. **HABITAT** forests, grassy slopes.

ABOVE: *Tree Pipit*

Red-fronted Serin ▪ *Serinus pusillus* 13cm

DESCRIPTION Sexes alike. Scarlet-orange forehead; blackish-grey crown; buffy back, streaked dark; yellow-orange rump and shoulder; yellow wing edges and whitish wing-bars; sooty-brown below, with grey and buff; dull yellow-buff belly and flanks, streaked brown.

Gregarious; quite active and constantly on the move; feeds on flower-heads and on ground; drinks and bathes often; spends considerable time in bushes and low trees. **FOOD** flower and grass seeds; small berries. **VOICE** pleasant twittering *chrr… chrr*, a faint *tree…tree…* call note. **DISTRIBUTION** W Himalaya, extreme west to Uttarakhand; 750–4,500m; breeds mostly at 2,400–4,000m. **HABITAT** rocky, bush-covered mountainsides.

European Goldfinch ■ *Carduelis carduelis* 14cm

DESCRIPTION Sexes alike. Crimson forehead; greyish-brown above, with large, white rump-patch; black and yellow wings striking, at rest and in flight. Young birds have streaked upperparts. Sociable; flock size ranges from four to several dozen together, sometimes along with other finches; forages on ground; also attends to flower-heads; undulating, somewhat dancing flight. **FOOD** seeds of flowers, especially thistle and sunflowers. **VOICE** ordinary call note a somewhat liquid *witwit...witwit...*; pleasant, twittering song; also a *chhrrik* call. **DISTRIBUTION** the Himalaya, extreme west to around C Nepal; breeds mostly 2,000–4,000m, ascending somewhat more; descends into foothills in winter. **HABITAT** open coniferous forests, orchards, cultivation, scrub.

Plain Mountain Finch ■ *Leucosticte nemoricola* 15cm

DESCRIPTION Sexes alike. Grey-brown above, streaked dark brown; greyer on rump; pale-buffy bar and markings in dark brown wings; dull grey-brown below, streaked browner on breast sides and flanks. **Brandt's Mountain Finch** *L. brandti* (18cm) is darker above, with rosy-pink rump and white in outer tail. The finches with plenty of white in their wings, and generally found in the high Tibetan country of the Himalaya, are generally snowfinches. Gregarious; good-sized flocks on ground, amidst stones; sometimes associates with other finches and buntings; calls often when feeding. **FOOD** grass and other seeds; small insects. **VOICE** twittering and chattering notes, rather sparrow-like in tone; calls frequently. **DISTRIBUTION** high Himalaya, breeds between 3,200–4,800m (above timberline); descends in winter, occasionally to as low as 1,000m. **HABITAT** open meadows, dwarf scrub, cultivation.

Brandt's Mountain Finch

Common Rosefinch ■ *Carpodacus erythrinus* 15cm

DESCRIPTION Male: crimson above, tinged brown; dark eye-stripe; crimson rump and underbody, fading into dull rose-white belly. Female: buff-brown above, streaked dark; two pale wing-bars; dull buff below, streaked, except on belly. The male **Pink-browed Rosefinch** *C. rodochroa* has a pink supercilium, rump and underparts. Small flocks; feeds on bushes and crops; often descends to ground; associates with other birds. **FOOD** crop seeds, fruit, buds, nectar. **VOICE** rather quiet in winter; pleasant song of up to 8 notes; may sing before departure from wintering grounds; also a double-noted questioning, *twee… ee* call. **DISTRIBUTION** breeds in the Himalaya, 2,700–4,000m; winters over most of India. **HABITAT** cultivation, open forests, gardens, bushes.

Pink-browed Rosefinch

White-browed Rosefinch ■ *Carpodacus thura* 17cm

DESCRIPTION Male: brown above, streaked blackish; pink and white forehead and supercilium; dark eye-stripe; rose-pink rump and double wing bar. Female: streaked brown; broad, whitish supercilium and single wing bar; yellow rump; buffy below, streaked. The white in supercilium easily identifies this species. Small flocks, either by themselves or with other finches; mostly feeds on ground, but settles on bushes and small trees. **FOOD** seeds, berries. **VOICE** calls often when feeding on ground, a fairly loud *pupuepipi…* call. **DISTRIBUTION** the Himalaya, breeds 3,000–4,000m; winters to about 1,800m. **HABITAT** tree-line forests, fir, juniper, rhododendron; open mountainsides and bushes in winter.

Red-headed Bullfinch

■ *Pyrrhula erythrocephala* 17cm

DESCRIPTION Male: black around base of beak and eye; brick-red crown; grey back; white rump; glossy purple-black wings; forked tail; black chin; rust-red below; ashy-white belly. Female: like male, but olive-yellow on crown; grey-brown back and underbody. The male **Orange Bullfinch** *P. aurantiaca* (14cm) of W Himalaya has orange-yellow back and underbody; female is yellow-brown. Small parties, occasionally with other birds; feed in low bushes, sometimes on ground; a bird of cover, rather quiet and secretive. **FOOD** seeds, buds, berries; also flower nectar. **VOICE** single or double-noted *pheu...pheu...* call. **DISTRIBUTION** the Himalaya, Kashmir to extreme east; breeds at 2,400–4,000m; descends in winter to about 1,200m. **HABITAT** forests, bushes.

Orange Bullfinch

Black-and-yellow Grosbeak ■ *Mycerobas icterioides* 22cm

DESCRIPTION Male: black head, throat, wings, tail and thighs; yellow collar, back and underbody below breast; thick, finch-like bill. Female: grey above; buffy rump and belly. The very similar male **Collared Grosbeak** *M. affinis* is brighter yellow (often with orangish wash), with yellow thighs. Small parties in tall coniferous forest; also feeds on ground and bushes, but spends much time in higher branches, where difficult to see; rather noisy. **FOOD** conifer seeds, shoots; also berries and insects. **VOICE** loud 2- or 3-noted whistle is familiar bird call of W Himalaya, loud *chuck...chuck* note when feeding; rich song of male. **DISTRIBUTION** W Himalaya up to C Nepal. **HABITAT** mountain forests.

Collared Grosbeak

Rock Bunting ■ *Emberiza cia* 15cm

DESCRIPTION Male: blue-grey head with black coronal stripe, eye-stripe and malar stripe, the latter curled and meeting eye-stripe diagnostic; whitish supercilium and cheeks; pale chestnut-brown back, streaked dark; unmarked rump; white outer sides of dark tail distinctive; blue-grey throat and breast; rufous-chestnut

below. Female: slightly duller. The male **Chestnut-eared Bunting** E. *fucata* has a black-streaked grey head, white throat, breast and chestnut ear-coverts. Solitary or in small parties; active and restless; mostly feeds on ground, meadows, paths and roads; flicks tail often; regularly settles on bushes and trees. **FOOD** seeds, small insects. **VOICE** squeaky *tsip... tsip...* note; calls often; common bird call of W Himalaya; has squeaky song of several notes. **DISTRIBUTION** the

entire Himalaya between 1,500–4,200m; most common in W Himalaya; winters in the foothills and plains of N India, coming as far south as Delhi. **HABITAT** grassy, rocky hillsides in open forests; cultivation, scrub.

RIGHT: *Chestnut-eared Bunting*

Grey-necked Bunting
■ *Emberiza buchanani* 15cm

DESCRIPTION Male: grey head with white eye-ring; brown back, with faint rufous wash and dark streaks; white edges to dark tail; whitish throat, mottled rufous; dark moustachial stripe, not easily visible; pale rufous-chestnut below. Female: somewhat duller than male; more prominent moustachial stripe. Winter visitor; small flocks; feeds mostly on ground, sometimes along with other birds; quite active. **FOOD** grass seeds, sometimes grain. **VOICE** a faint single note. **DISTRIBUTION** winter visitor; quite common over W and C India-Gujarat, S and W Rajasthan, SW Uttar Pradesh, south through W and C Madhya Pradesh, Maharashtra and parts of Karnataka. **HABITAT** open, rocky grassy country, scrub.

White-capped Bunting
▪ *Emberiza stewarti* 15cm

DESCRIPTION Male: grey-white top of head; black eye-stripe, whitish cheeks, black chin and upper throat distinctive; chestnut back and rump; white outer tail; white breast with chestnut gorget below; dull fulvous below, chestnut flanks. Female: lacks black and white head pattern of male; brown above, streaked; rufous-chestnut rump; fulvous-buff below, with rufous breast. The male **Striolated Bunting** E. *striolata* (14cm), with more or less overlapping range, has grey-white head, completely streaked black. Small flocks, often with other buntings and finches; feeds on ground; rests in bushes and trees. **FOOD** chiefly grass seeds. **VOICE** faint but sharp *tsit...* or *chit...* note. **DISTRIBUTION** breeds in W Himalaya, extreme west to Uttarakhand, 1,500–3,500m; winters in W Himalayan foothills, and over extensive parts of W and C India, south to Maharashtra. **HABITAT** open, grass-covered, rocky hillsides, scrub.

Striolated Bunting

Crested Bunting ▪ *Melophus lathami* 15cm

DESCRIPTION Male: striking glossy black plumage, with long, pointed crest and chestnut wings and tail. Female: crested; olive-brown above, streaked darker, rufous in wings distinctive; buffy-yellow below, streaked dark on breast; darkish moustachial stripe. Small flocks, often spread wide over an area; feeds on ground, on paths, meadows and tar roads, especially along mountainsides; perches on ruins, walls, stones and low bushes; on ground, an active and upright bird. **FOOD** grass seeds; presumably also insects. **VOICE** faint *chip...* call; pleasant, though somewhat monotonous song of breeding male (May–August). **DISTRIBUTION** resident over wide part of India, from outer Himalaya to about 1,800m; south to SW Maharashtra and N Andhra Pradesh; appears to move considerably after the rains. **HABITAT** open, bush and rock-covered mountainsides, open country; sometimes also cultivation.

STATUS
R   widespread resident
r   very local resident
W   widespread winter visitor
w   sparse winter visitor
P   widespread migrant
p   sparse migrant
V   vagrant or irregular visitor
s   local summer breeder
I   introduced resident

IUCN RED LIST STATUS
(Ex)   extinct
?   requires status confirmation
CR   critically endangered
EN   endangered
DD   conservation dependant
NT   near-threatened
LC   Least concern
VU   Vulnerable

Common English Name	Scientific Name	Status	IUCN
Megapodiidae (Megapode)			
Nicobar Megapode	*Megapodius nicobariensis*	r	VU
Phasianidae (Partridges and Pheasants)			
Snow Partridge	*Lerwa lerwa*	V	LC
Tibetan Snowcock	*Tetraogallus tibetanus*	r	LC
Himalayan Snowcock	*Tetraogallus himalayensis*	r	LC
Buff-throated Partridge	*Tetraophasis szechenyii*	?	LC
Chukar Partridge	*Alectoris chukar*	R	LC
See-See Partridge	*Ammoperdix griseogularis*	r	LC
Black Francolin	*Francolinus francolinus*	R	LC
Painted Francolin	*Francolinus pictus*	R	LC
Chinese Francolin	*Francolinus pintadeanus*	r	LC
Grey Francolin	*Francolinus pondicerianus*	R	LC
Swamp Francolin	*Francolinus gularis*	r	VU
Tibetan Partridge	*Perdix hodgsoniae*	r	LC
Common Quail	*Coturnix coturnix*	rw	LC
Japanese Quail	*Coturnix japonica*	w	NT
Rain Quail	*Coturnix coromandelica*	r	LC
King Quail	*Coturnix chinensis*	r	LC
Jungle Bush Quail	*Perdicula asiatica*	R	LC
Rock Bush Quail	*Perdicula argoondah*	R	LC
Painted Bush Quail	*Perdicula erythrorhyncha*	r	LC
Himalayan Quail	*Ophrysia superciliosa*	Ex?	CR
Manipur Bush Quail	*Perdicula manipurensis*	R	VU
Hill Partridge	*Arborophila torqueola*	r	LC
Rufous-throated Partridge	*Arborophila rufogularis*	r	LC
White-cheeked Partridge	*Arborophila atrogularis*	r	NT
Chestnut-breasted Partridge	*Arborophila mandellii*	r	VU
Mountain Bamboo Partridge	*Bambusicola fytchii*	r	LC
Red Spurfowl	*Galloperdix spadicea*	r	LC
Sri Lanka Spurfowl	*Galloperdix bicalcarata*	r	LC
Painted Spurfowl	*Galloperdix lunulata*	r	LC
Blood Pheasant	*Ithaginis cruentus*	r	LC
Western Tragopan	*Tragopan melanocephalus*	r	VU
Satyr Tragopan	*Tragopan satyra*	r	NT
Blyth's Tragopan	*Tragopan blythii*	r	VU
Temminck's Tragopan	*Tragopan temminckii*	r	LC
Koklass Pheasant	*Pucrasia macrolopha*	r	LC
Himalayan Monal	*Lophophorus impejanus*	r	LC
Sclater's Monal	*Lophophorus sclateri*	r	VU
Red Junglefowl	*Gallus gallus*	R	LC
Sri Lanka Junglefowl	*Gallus lafayetii*	R	LC
Grey Junglefowl	*Gallus sonneratii*	R	LC
Kalij Pheasant	*Lophura leucomelanos*	R	LC
Tibetan Eared Pheasant	*Crossoptilon harmani*	r	NT
Cheer Pheasant	*Catreus wallichii*	r	VU
Mrs Hume's Pheasant	*Syrmaticus humiae*	r	NT
Grey Peacock Pheasant	*Polyplectron bicalcaratum*	r	LC
Indian Peafowl	*Pavo cristatus*	R	LC
Green Peafowl	*Pavo muticus*	r	EN
Anatidae (Ducks, Geese and Swans)			
Fulvous Whistling-duck	*Dendrocygna bicolor*	r	LC
Lesser Whistling-duck	*Dendrocygna javanica*	R	LC
Bean Goose	*Anser fabalis*	V	LC
Greater White-fronted Goose	*Anser albifrons*	V	LC
Lesser White-fronted Goose	*Anser erythropus*	V	VU

Common English Name	Scientific Name	Status	IUCN
Greylag Goose	Anser anser	W	LC
Bar-headed Goose	Anser indicus	rw	LC
Snow Goose	Anser caerulescens	?	LC
Red-breasted Goose	Branta ruficollis	?	LC
Mute Swan	Cygnus olor	V	LC
Whooper Swan	Cygnus cygnus	V	LC
Tundra Swan	Cygnus columbianus	V	LC
Knob-billed Duck	Sarkidiornis melanotos	r	LC
Ruddy Shelduck	Tadorna ferruginea	RW	LC
Common Shelduck	Tadorna tadorna	w	LC
White-winged Duck	Asarcornis scutulata	r	EN
Cotton Pygmy-goose	Nettapus coromandelianus	r	LC
Mandarin Duck	Aix galericulata	V	LC
Gadwall	Anas strepera	W	LC
Falcated Duck	Anas falcata	V	NT
Eurasian Wigeon	Anas penelope	W	LC
Mallard	Anas platyrhynchos	rW	LC
Indian Spot-billed Duck	Anas poecilorhyncha	R	LC
Eastern Spot-billed Duck	Anas zonorhyncha	r	LC
Northern Shoveler	Anas clypeata	W	LC
Sunda Teal	Anas gibberifrons	r	LC
Northern Pintail	Anas acuta	W	LC
Garganey	Anas querquedula	W	LC
Baikal Teal	Anas formosa	V	LC
Common Teal	Anas crecca	W	LC
Marbled Duck	Marmaronetta angustirostris	V	VU
Red-crested Pochard	Netta rufina	w	LC
Common Pochard	Aythya ferina	W	LC
Ferruginous Duck	Aythya nyroca	w	NT
Baer's Pochard	Aythya baeri	w	CR
Tufted Duck	Aythya fuligula	W	LC
Greater Scaup	Aythya marila	V	LC
Long-tailed Duck	Clangula hyemalis	V	VU
Common Goldeneye	Bucephala clangula	V	LC
Smew	Mergellus albellus	w	LC
Goosander	Mergus merganser	RW	LC
Red-breasted Merganser	Mergus serrator	V	LC
White-headed Duck	Oxyura leucocephala	V	EN
Pink-headed Duck	Rhodonessa caryophyllacea	Ex?	CR
Gaviidae (Divers)			
Red-throated Diver	Gavia stellata	V	LC
Black-throated Diver	Gavia arctica	V	LC
Procellariidae (Shearwaters and Petrels)			
Cape Petrel	Daption capense	V	LC
Barau's Petrel	Pterodroma baraui	V	LC
Bulwer's Petrel	Bulweria bulwerii	V	LC
Jouanin's Petrel	Bulweria fallax	V	LC
Streaked Shearwater	Calonectris leucomelas	V	LC
Wedge-tailed Shearwater	Puffinus pacificus	V	LC
Flesh-footed Shearwater	Puffinus carneipes	s	LC
Audubon's Shearwater	Puffinus lherminieri	r	LC
Sooty Shearwater	Puffinus griseus	V	LC
Short-tailed Shearwater	Puffinus tenuirostris	V	LC
Persian Shearwater	Puffinus persicus	P	NT
Hydrobatidae (Storm Petrels)			
Wilson's Storm-petrel	Oceanites oceanicus	P	LC
White-faced Storm-petrel	Pelagodroma marina	V	LC
Black-bellied Storm-petrel	Fregetta tropica	V	LC
Swinhoe's Storm-petrel	Oceanodroma monorhis	V	LC
Podicipedidae (Grebes)			
Little Grebe	Tachybaptus ruficollis	R	LC
Red-necked Grebe	Podiceps grisegena	W	LC
Great Crested Grebe	Podiceps cristatus	rw	LC
Slavonian Grebe	Podiceps auritus	w	LC
Black-necked Grebe	Podiceps nigricollis	rw	LC
Phoenicopteridae (Flamingos)			
Greater Flamingo	Phoenicopterus roseus	rW	LC
Lesser Flamingo	Phoenicopterus minor	r	NT
Phaethontidae (Tropicbirds)			
Red-billed Tropicbird	Phaethon aethereus	w	LC
Red-tailed Tropicbird	Phaethon rubricauda	r	LC
White-tailed Tropicbird	Phaethon lepturus	r	LC
Ardeidae (Bitterns, Herons and Egrets)			
Little Egret	Egretta garzetta	R	LC

Common English Name	Scientific Name	Status	IUCN
Western Reef Egret	Egretta gularis	r	LC
Pacific Reef Egret	Egretta sacra	r	LC
Grey Heron	Ardea cinerea	RW	LC
Goliath Heron	Ardea goliath	r	LC
White-bellied Heron	Ardea insignis	r	CR
Purple Heron	Ardea purpurea	R	LC
Great Egret	Casmerodius albus	RW	LC
Intermediate Egret	Mesophoyx intermedia	R	LC
Cattle Egret	Bubulcus ibis	R	LC
Indian Pond Heron	Ardeola grayii	R	LC
Chinese Pond Heron	Ardeola bacchus	r	LC
Striated Heron	Butorides striata	r	LC
Black-crowned Night Heron	Nycticorax nycticorax	R	LC
Malayan Night Heron	Gorsachius melanolophus	r	LC
Little Bittern	Ixobrychus minutus	r	LC
Yellow Bittern	Ixobrychus sinensis	r	LC
Cinnamon Bittern	Ixobrychus cinnamomeus	r	LC
Black Bittern	Dupetor flavicollis	r	LC
Great Bittern	Botaurus stellaris	w	LC
Ciconiidae (Storks)			
Painted Stork	Mycteria leucocephala	R	NT
Asian Openbill	Anastomus oscitans	R	LC
Black Stork	Ciconia nigra	w	LC
Woolly-necked Stork	Ciconia episcopus	R	LC
White Stork	Ciconia ciconia	w	LC
Black-necked Stork	Ephippiorhynchus asiaticus	r	NT
Lesser Adjutant	Leptoptilos javanicus	r	VU
Greater Adjutant	Leptoptilos dubius	r	EN
Threskiornithidae (Ibises and Spoonbill)			
Glossy Ibis	Plegadis falcinellus	RW	LC
Black-headed Ibis	Threskiornis melanocephalus	R	NT
Red-naped Ibis	Pseudibis papillosa	R	LC
Eurasian Spoonbill	Platalea leucorodia	RW	LC
Pelecanidae (Pelicans)			
Great White Pelican	Pelecanus onocrotalus	rW	LC
Dalmatian Pelican	Pelecanus crispus	r	VU
Spot-billed Pelican	Pelecanus philippensis	R	NT
Fregatidae (Frigatebirds)			
Great Frigatebird	Fregata minor	P	LC
Lesser Frigatebird	Fregata ariel	r	LC
Christmas Island Frigatebird	Fregata andrewsi	V	CR
Sulidae (Boobies)			
Masked Booby	Sula dactylatra	r	LC
Red-footed Booby	Sula sula	r	LC
Brown Booby	Sula leucogaster	r	LC
Phalacrocoracidae (Cormorants)			
Little Cormorant	Phalacrocorax niger	R	LC
Indian Cormorant	Phalacrocorax fuscicollis	R	LC
Great Cormorant	Phalacrocorax carbo	RW	LC
Pygmy Cormorant	Phalacrocorax pygmeus	r	LC
Anhingidae (Darter)			
Darter	Anhinga melanogaster	R	NT
Pandionidae (Osprey)			
Osprey	Pandion haliaetus	rW	LC
Accipitridae (Hawks, Kites and Eagles)			
Jerdon's Baza	Aviceda jerdoni	r	LC
Black Baza	Aviceda leuphotes	r	LC
Oriental Honey-buzzard	Pernis ptilorhyncus	RW	LC
Black-winged Kite	Elanus caeruleus	R	LC
Red Kite	Milvus milvus	?	LC
Black Kite	Milvus migrans	RW	LC
Black-eared Kite	Milvus m. lineatus	W	LC
Brahminy Kite	Haliastur indus	R	LC
White-bellied Sea Eagle	Haliaeetus leucogaster	R	LC
Pallas's Fish Eagle	Haliaeetus leucoryphus	r	VU
White-tailed Eagle	Haliaeetus albicilla	w	LC
Lesser Fish Eagle	Ichthyophaga humilis	r	NT
Grey-headed Fish Eagle	Ichthyophaga ichthyaetus	r	NT
Bearded Vulture	Gypaetus barbatus	r	LC
Egyptian Vulture	Neophron percnopterus	R	EN
White-rumped Vulture	Gyps bengalensis	r	CR
Indian Vulture	Gyps indicus	r	CR
Slender-billed Vulture	Gyps tenuirostris	r	CR
Himalayan Vulture	Gyps himalayensis	r	LC
Griffon Vulture	Gyps fulvus	r	LC

Common English Name	Scientific Name	Status	IUCN
Cinereous Vulture	Aegypius monachus	rw	NT
Red-headed Vulture	Sarcogyps calvus	r	CR
Short-toed Snake Eagle	Circaetus gallicus	r	LC
Crested Serpent Eagle	Spilornis cheela	R	LC
Great Nicobar Serpent Eagle	Spilornis klossi	r	NT
Central Nicobar Serpent Eagle	Spilornis minimus	r	NT
Andaman Serpent Eagle	Spilornis elgini	r	NT
Eurasian Marsh Harrier	Circus aeruginosus	W	LC
Eastern Marsh Harrier	Circus a. spilonotus	V	LC
Hen Harrier	Circus cyaneus	w	LC
Pallid Harrier	Circus macrourus	w	NT
Pied Harrier	Circus melanoleucos	rw	LC
Montagu's Harrier	Circus pygargus	w	LC
Crested Goshawk	Accipiter trivirgatus	w	LC
Shikra	Accipiter badius	R	LC
Nicobar Sparrowhawk	Accipiter butleri	w	VU
Chinese Sparrowhawk	Accipiter soloensis	W	LC
Japanese Sparrowhawk	Accipiter gularis	w	LC
Besra	Accipiter virgatus	r	LC
Eurasian Sparrowhawk	Accipiter nisus	rw	LC
Northern Goshawk	Accipiter gentilis	rw	LC
White-eyed Buzzard	Butastur teesa	R	LC
Common Buzzard	Buteo buteo	rw	LC
Himalayan Buzzard	Buteo b. burmanicus	rw	LC
Long-legged Buzzard	Buteo rufinus	rW	LC
Upland Buzzard	Buteo hemilasius	w	LC
Black Eagle	Ictinaetus malayensis	r	LC
Indian Spotted Eagle	Aquila hastata	r	VU
Greater Spotted Eagle	Aquila clanga	W	VU
Tawny Eagle	Aquila rapax	R	LC
Steppe Eagle	Aquila nipalensis	W	LC
Eastern Imperial Eagle	Aquila heliaca	w	VU
Golden Eagle	Aquila chrysaetos	r	LC
Bonelli's Eagle	Aquila fasciata	r	LC
Booted Eagle	Hieraaetus pennatus	w	LC
Rufous-bellied Eagle	Lophotriorchis kienerii	r	LC
Crested Hawk Eagle	Nisaetus cirrhatus	R	LC
Changeable Hawk Eagle	Nisaetus c. limnaeetus	R	LC
Mountain Hawk Eagle	Nisaetus nipalensis	r	LC
Legge's Hawk Eagle	Nisaetus n. keelarti	r	LC
Falconidae (Falcons)			
Collared Falconet	Microhierax caerulescens	r	LC
Pied Falconet	Microhierax melanoleucus	r	LC
Lesser Kestrel	Falco naumanni	p	LC
Common Kestrel	Falco tinnunculus	RW	LC
Red-necked Falcon	Falco chicquera	r	LC
Amur Falcon	Falco amurensis	p	LC
Merlin	Falco columbarius	w	LC
Sooty Falcon	Falco concolor	r	LC
Eurasian Hobby	Falco subbuteo	rp	LC
Oriental Hobby	Falco severus	r	LC
Laggar Falcon	Falco jugger	r	NT
Saker Falcon	Falco cherrug	w	EN
Peregrine Falcon	Falco peregrinus	rw	LC
Barbary Falcon	Falco p. pelegrinoides	rw	LC
Otididae (Bustards)			
Little Bustard	Tetrax tetrax	V	NT
Great Bustard	Otis tarda	V	VU
Great Indian Bustard	Ardeotis nigriceps	r	CR
Macqueen's Bustard	Chlamydotis macqueenii	w	VU
Bengal Florican	Houbaropsis bengalensis	r	CR
Lesser Florican	Sypheotides indicus	r	EN
Heliornithidae (Finfoot)			
Masked Finfoot	Heliopais personatus	r	EN
Rallidae (Crakes, Rails)			
Andaman Crake	Rallina canningi	r	NT
Slaty-legged Crake	Rallina eurizonoides	r	LC
Slaty-breasted Rail	Gallirallus striatus	r	LC
Water Rail	Rallus aquaticus	rw	LC
Brown-cheeked Rail	Rallus indicus	r	LC
Corn Crake	Crex crex	V	LC
Brown Crake	Amaurornis akool	r	LC
White-breasted Waterhen	Amaurornis phoenicurus	R	LC
Black-tailed Crake	Porzana bicolor	r	LC

Common English Name	Scientific Name	Status	IUCN
Little Crake	Porzana parva	V	LC
Baillon's Crake	Porzana pusilla	rw	LC
Spotted Crake	Porzana porzana	V	LC
Ruddy-breasted Crake	Porzana fusca	r	LC
Watercock	Gallicrex cinerea	r	LC
Purple Swamphen	Porphyrio porphyrio	R	LC
Common Moorhen	Gallinula chloropus	R	LC
Eurasian Coot	Fulica atra	RW	LC
Gruidae (Cranes)			
Siberian Crane	Grus leucogeranus	?	CR
Sarus Crane	Grus antigone	r	VU
Demoiselle Crane	Grus virgo	w	LC
Common Crane	Grus grus	w	LC
Black-necked Crane	Grus nigricollis	r	VU
Turnicidae (Buttonquails)			
Small Buttonquail	Turnix sylvaticus	R	LC
Yellow-legged Buttonquail	Turnix tanki	R	LC
Barred Buttonquail	Turnix suscitator	R	LC
Burhinidae (Thick-knees)			
Eurasian Thick-knee	Burhinus oedicnemus	?	LC
Indian Thick-knee	Burhinus o. indicus	R	LC
Great Thick-knee	Esacus recurvirostris	r	LC
Beach Thick-knee	Esacus neglectus	r	NT
Haematopodidae (Oystercatchers)			
Eurasian Oystercatcher	Haematopus ostralegus	w	LC
Dromadidae (Crab Plover)			
Crab Plover	Dromas ardeola	w	LC
Ibidorhynchidae (Ibisbill)			
Ibisbill	Ibidorhyncha struthersii	r	LC
Recurvirostridae (Stilts and Avocets)			
Black-winged Stilt	Himantopus himantopus	RW	LC
Pied Avocet	Recurvirostra avosetta	rW	LC
Charadriidae (Plovers and Lapwings)			
European Golden Plover	Pluvialis apricaria	V	LC
Pacific Golden Plover	Pluvialis fulva	W	LC
Grey Plover	Pluvialis squatarola	w	LC
Common Ringed Plover	Charadrius hiaticula	w	LC
Long-billed Plover	Charadrius placidus	w	LC
Little Ringed Plover	Charadrius dubius	RW	LC
Kentish Plover	Charadrius alexandrinus	RW	LC
Lesser Sand Plover	Charadrius mongolus	sW	LC
Greater Sand Plover	Charadrius leschenaultii	w	LC
Caspian Plover	Charadrius asiaticus	w	LC
Oriental Plover	Charadrius veredus	w	LC
Northern Lapwing	Vanellus vanellus	w	LC
Yellow-wattled Lapwing	Vanellus malabaricus	w	LC
River Lapwing	Vanellus duvaucelii	R	NT
Grey-headed Lapwing	Vanellus cinereus	w	LC
Red-wattled Lapwing	Vanellus indicus	R	LC
Sociable Lapwing	Vanellus gregarius	w	CR
White-tailed Lapwing	Vanellus leucurus	W	LC
Rostratulidae (Painted-snipe)			
Greater Painted-snipe	Rostratula benghalensis	r	LC
Jacanidae (Jacanas)			
Pheasant-tailed Jacana	Hydrophasianus chirurgus	R	LC
Bronze-winged Jacana	Metopidius indicus	R	LC
Scolopacidae (Snipes, Sandpipers and other Waders)			
Eurasian Woodcock	Scolopax rusticola	rW	LC
Solitary Snipe	Gallinago solitaria	r	LC
Wood Snipe	Gallinago nemoricola	r	VU
Pintail Snipe	Gallinago stenura	W	LC
Swinhoe's Snipe	Gallinago megala	w	LC
Great Snipe	Gallinago media	V	LC
Common Snipe	Gallinago gallinago	rW	LC
Jack Snipe	Lymnocryptes minimus	w	LC
Black-tailed Godwit	Limosa limosa	W	NT
Eastern Black-tailed Godwit	Limosa l. melanuroides	w	NT
Bar-tailed Godwit	Limosa lapponica	W	LC
Whimbrel	Numenius phaeopus	W	LC
Eurasian Curlew	Numenius arquata	W	NT
Eastern Curlew	Numenius madagascariensis	w	NT
Spotted Redshank	Tringa erythropus	W	LC
Common Redshank	Tringa totanus	sW	LC
Marsh Sandpiper	Tringa stagnatilis	W	LC

Common English Name	Scientific Name	Status	IUCN
Common Greenshank	*Tringa nebularia*	W	LC
Nordmann's Greenshank	*Tringa guttifer*	?	EN
Green Sandpiper	*Tringa ochropus*	W	LC
Wood Sandpiper	*Tringa glareola*	W	LC
Grey-tailed Tattler	*Tringa brevipes*	V	LC
Terek Sandpiper	*Xenus cinereus*	w	LC
Common Sandpiper	*Actitis hypoleucos*	sW	LC
Ruddy Turnstone	*Arenaria interpres*	w	LC
Long-billed Dowitcher	*Limnodromus scolopaceus*	V	LC
Asian Dowitcher	*Limnodromus semipalmatus*	w	NT
Great Knot	*Calidris tenuirostris*	w	VU
Red Knot	*Calidris canutus*	W	LC
Sanderling	*Calidris alba*	w	LC
Little Stint	*Calidris minuta*	W	LC
Red-necked Stint	*Calidris ruficollis*	w	LC
Temminck's Stint	*Calidris temminckii*	W	LC
Long-toed Stint	*Calidris subminuta*	w	LC
Sharp-tailed Sandpiper	*Calidris acuminate*	V	LC
Pectoral Sandpiper	*Calidris melanotos*	V	LC
Dunlin	*Calidris alpina*	w	LC
Curlew Sandpiper	*Calidris ferruginea*	W	LC
Spoon-billed Sandpiper	*Eurynorhynchus pygmeus*	w	VU
Buff-breasted Sandpiper	*Tryngites subruficollis*	V	LC
Broad-billed Sandpiper	*Limicola falcinellus*	w	LC
Ruff	*Philomachus pugnax*	W	LC
Red-necked Phalarope	*Phalaropus lobatus*	w	LC
Red Phalarope	*Phalaropus fulicarius*	V	LC
Glareolidae (Coursers & Pratincoles)			
Jerdon's Courser	*Rhinoptilus bitorquatus*	r	CR
Cream-colored Courser	*Cursorius cursor*	rw	LC
Indian Courser	*Cursorius coromandelicus*	r	LC
Collared Pratincole	*Glareola pratincola*	rw	LC
Oriental Pratincole	*Glareola maldivarum*	R	LC
Small Pratincole	*Glareola lactea*	R	LC
Laridae (Gulls, Terns, Noddies & Skimmer)			
Caspian Gull	*Larus cachinnans*	w	LC
Heuglin's Gull	*Larus heuglini*	w	LC
Steppe Gull	*Larus h. barabensis*	W	LC
Mew Gull	*Larus canus*	V	LC
Pallas's Gull	*Ichthyaetus ichthyaetus*	w	LC
Sooty Gull	*Ichthyaetus hemprichii*	V	LC
Brown-headed Gull	*Chroicocephalus brunnicephalus*	sW	LC
Black-headed Gull	*Chroicocephalus ridibundus*	R	LC
Slender-billed Gull	*Chroicocephalus genei*	w	LC
Little Gull	*Hydrocoloeus minutus*	V	LC
Black-legged Kittiwake	*Rissa tridactyla*	V	LC
Gull-billed Tern	*Gelochelidon nilotica*	rW	LC
Caspian Tern	*Hydroprogne caspia*	W	LC
Lesser Crested Tern	*Thalasseus bengalensis*	r	LC
Greater Crested Tern	*Thalasseus bergii*	L	LC
Sandwich Tern	*Thalasseus sandvicensis*	w	LC
River Tern	*Sterna aurantia*	R	NT
Roseate Tern	*Sterna dougallii*	r	LC
Black-naped Tern	*Sterna sumatrana*	r	LC
Common Tern	*Sterna hirundo*	sW	LC
Arctic Tern	*Sterna paradisaea*	V	LC
White-cheeked Tern	*Sterna repressa*	p	LC
Black-bellied Tern	*Sterna acuticauda*	r	EN
Little Tern	*Sternula albifrons*	r	LC
Saunders's Tern	*Sternula saundersi*	r	LC
Bridled Tern	*Onychoprion anaethetus*	s	LC
Sooty Tern	*Onychoprion fuscatus*	s	LC
Whiskered Tern	*Chlidonias hybrida*	RW	LC
White-winged Tern	*Chlidonias leucopterus*	w	LC
Black Tern	*Chlidonias niger*	V	LC
White Tern	*Gygis alba*	V	LC
Brown Noddy	*Anous stolidus*	V	LC
Black Noddy	*Anous minutus*	V	LC
Lesser Noddy	*Anous tenuirostris*	V	LC
Indian Skimmer	*Rynchops albicollis*	r	VU
Stercorariidae (Skuas)			
Brown Skua	*Stercorarius antarcticus*	p	LC
South Polar Skua	*Stercorarius maccormicki*	V	LC
Long-tailed Skua	*Stercorarius longicaudus*	?	LC

Common English Name	Scientific Name	Status	IUCN
Pomarine Skua	*Stercorarius pomarinus*	p	LC
Arctic Skua	*Stercorarius parasiticus*	p	LC
Pteroclidae (Sandgrouse)			
Tibetan Sandgrouse	*Syrrhaptes tibetanus*	r	LC
Pallas's Sandgrouse	*Syrrhaptes paradoxus*	V	LC
Pin-tailed Sandgrouse	*Pterocles alchata*	w	LC
Chestnut-bellied Sandgrouse	*Pterocles exustus*	r	LC
Spotted Sandgrouse	*Pterocles senegallus*	w	LC
Black-bellied Sandgrouse	*Pterocles orientalis*	rw	LC
Crowned Sandgrouse	*Pterocles coronatus*	r	LC
Lichtenstein's Sandgrouse	*Pterocles lichtensteinii*	r	LC
Crowned Sandgrouse	*Pterocles coronatus*	r	LC
Painted Sandgrouse	*Pterocles indicus*	r	LC
Columbidae (Pigeons and Doves)			
Common Pigeon	*Columba livia*	R	LC
Hill Pigeon	*Columba rupestris*	R	LC
Snow Pigeon	*Columba leuconota*	R	LC
Yellow-eyed Pigeon	*Columba eversmanni*	w	VU
Common Wood Pigeon	*Columba palumbus*	w	LC
Speckled Wood Pigeon	*Columba hodgsonii*	r	LC
Ashy Wood Pigeon	*Columba pulchricollis*	r	LC
Nilgiri Wood Pigeon	*Columba elphinstonii*	r	VU
Pale-capped Pigeon	*Columba punicea*	w	VU
Sri Lanka Wood Pigeon	*Columba torringtoniae*	r	VU
Andaman Wood Pigeon	*Columba palumboides*	r	NT
European Turtle Dove	*Streptopelia turtur*	V	LC
Oriental Turtle Dove	*Streptopelia orientalis*	RW	LC
Red Collared Dove	*Streptopelia tranquebarica*	R	LC
Eurasian Collared Dove	*Streptopelia decaocto*	R	LC
Laughing Dove	*Stigmatopelia senegalensis*	R	LC
Spotted Dove	*Stigmatopelia chinensis*	R	LC
Barred Cuckoo Dove	*Macropygia unchall*	r	LC
Andaman Cuckoo Dove	*Macropygia rufipennis*	r	NT
Emerald Dove	*Chalcophaps indica*	R	LC
Nicobar Pigeon	*Caloenas nicobarica*	r	NT
Orange-breasted Green Pigeon	*Treron bicinctus*	r	LC
Sri Lanka Green Pigeon	*Treron pompadora*	r	LC
Grey-fronted Green Pigeon	*Treron p. affinis*	r	LC
Andaman Green Pigeon	*Treron p. chloropterus*	r	LC
Ashy-headed Green Pigeon	*Treron p. phayrei*	r	LC
Thick-billed Green Pigeon	*Treron curvirostra*	r	LC
Yellow-footed Green Pigeon	*Treron phoenicopterus*	R	LC
Pin-tailed Green Pigeon	*Treron apicauda*	r	LC
Wedge-tailed Green Pigeon	*Treron sphenurus*	r	LC
Green Imperial Pigeon	*Ducula aenea*	r	LC
Nicobar Imperial Pigeon	*Ducula a. nicobarica*	r	LC
Mountain Imperial Pigeon	*Ducula badia*	r	LC
Pied Imperial Pigeon	*Ducula bicolor*	r	LC
Psittacidae (Parrots and Parakeets)			
Vernal Hanging Parrot	*Loriculus vernalis*	R	LC
Sri Lanka Hanging Parrot	*Loriculus beryllinus*	r	LC
Alexandrine Parakeet	*Psittacula eupatria*	R	LC
Rose-ringed Parakeet	*Psittacula krameri*	R	LC
Slaty-headed Parakeet	*Psittacula himalayana*	R	LC
Grey-headed Parakeet	*Psittacula finschii*	R	LC
Plum-headed Parakeet	*Psittacula cyanocephala*	R	LC
Blossom-headed Parakeet	*Psittacula roseata*	R	LC
Malabar Parakeet	*Psittacula columboides*	R	LC
Derbyan Parakeet	*Psittacula derbiana*	r	NT
Red-breasted Parakeet	*Psittacula alexandri*	R	LC
Nicobar Parakeet	*Psittacula caniceps*	r	NT
Layard's Parakeet	*Psittacula calthropae*	r	LC
Long-tailed Parakeet	*Psittacula longicauda*	R	NT
Cuculidae (Cuckoos)			
Jacobin Cuckoo	*Clamator jacobinus*	rS	LC
Chestnut-winged Cuckoo	*Clamator coromandus*	r	LC
Large Hawk Cuckoo	*Hierococcyx sparverioides*	r	LC
Common Hawk Cuckoo	*Hierococcyx varius*	R	LC
Hodgson's Hawk Cuckoo	*Hierococcyx fugax*	r	LC
Indian Cuckoo	*Cuculus micropterus*	R	LC
Eurasian Cuckoo	*Cuculus canorus*	R	LC
Himalayan Cuckoo	*Cuculus saturatus*	r	LC
Lesser Cuckoo	*Cuculus poliocephalus*	r	LC
Banded Bay Cuckoo	*Cacomantis sonneratii*	r	LC

Common English Name	Scientific Name	Status	IUCN
Grey-bellied Cuckoo	Cacomantis passerinus	r	LC
Plaintive Cuckoo	Cacomantis merulinus	r	LC
Asian Emerald Cuckoo	Chrysococcyx maculatus	r	LC
Violet Cuckoo	Chrysococcyx xanthorhynchus	r	LC
Drongo Cuckoo	Surniculus lugubris	r	LC
Asian Koel	Eudynamys scolopaceus	R	LC
Green-billed Malkoha	Rhopodytes tristis	r	LC
Blue-faced Malkoha	Rhopodytes viridirostris	r	LC
Sirkeer Malkoha	Taccocua leschenaultii	r	LC
Red-faced Malkoha	Phaenicophaeus pyrrhocephalus	r	VU
Greater Coucal	Centropus sinensis	R	LC
Southern Coucal	Centropus s. parroti	R	LC
Brown Coucal	Centropus andamanensis	r	LC
Lesser Coucal	Centropus bengalensis	t	LC
Green-billed Coucal	Centropus chlorhynchus	r	VU
Tytonidae (Barn Owls)			
Barn Owl	Tyto alba	r	LC
Andaman Barn Owl	Tyto a. deroepstorffi	r	LC
Eastern Grass Owl	Tyto longimembris	r	LC
Oriental Bay Owl	Phodilus badius	r	LC
Sri Lanka Bay Owl	Phodilus b. assimilis	r	LC
Strigidae (Owls)			
Serendib Scops Owl	Otus thilohoffmanni	r	NT
Andaman Scops Owl	Otus balli	r	NT
Mountain Scops Owl	Otus spilocephalus	r	LC
Pallid Scops Owl	Otus brucei	r	LC
Eurasian Scops Owl	Otus scops	w	LC
Indian Scops Owl	Otus bakkamoena	!	LC
Collared Scops Owl	Otus b. lettia	R	LC
Oriental Scops Owl	Otus sunia	R	LC
Nicobar Scops Owl	Otus alius	r	LC
Eurasian Eagle Owl	Bubo bubo	R	LC
Indian Eagle Owl	Bubo b. bengalensis	R	LC
Spot-bellied Eagle Owl	Bubo nipalensis	r	LC
Dusky Eagle Owl	Bubo coromandus	R	LC
Brown Fish Owl	Ketupa zeylonensis	r	LC
Tawny Fish Owl	Ketupa flavipes	r	LC
Buffy Fish Owl	Ketupa ketupu	t	LC
Mottled Wood Owl	Strix ocellata	r	LC
Brown Wood Owl	Strix leptogrammica	r	LC
Tawny Owl	Strix aluco	r	LC
Himalayan Wood Owl	Strix a. nivicola	r	LC
Collared Owlet	Glaucidium brodiei	r	LC
Asian Barred Owlet	Glaucidium cuculoides	r	LC
Jungle Owlet	Glaucidium radiatum	R	LC
Chestnut-backed Owlet	Glaucidium castanotum	r	NT
Little Owl	Athene noctua	r	LC
Spotted Owlet	Athene brama	R	LC
Forest Owlet	Heteroglaux blewitti	r	CR
Boreal Owl	Aegolius funereus	V	LC
Brown Hawk Owl	Ninox scutulata	r	LC
Hume's Hawk Owl	Ninox s. obscura	r	NT
Andaman Hawk Owl	Ninox affinis	r	NT
Long-eared Owl	Asio otus	rw	LC
Short-eared Owl	Asio flammeus	w	LC
Podargidae (Frogmouths)			
Sri Lanka Frogmouth	Batrachostomus moniliger	r	LC
Hodgson's Frogmouth	Batrachostomus hodgsoni	r	LC
Caprimulgidae (Nightjars)			
Great Eared Nightjar	Eurostopodus macrotis	r	LC
Jungle Nightjar	Caprimulgus indicus	R	LC
Grey Nightjar	Caprimulgus i. jotaka	R	LC
European Nightjar	Caprimulgus europaeus	rp	LC
Egyptian Nightjar	Caprimulgus aegyptius	s	LC
Sykes's Nightjar	Caprimulgus mahrattensis	r	LC
Large-tailed Nightjar	Caprimulgus macrurus	R	LC
Jerdon's Nightjar	Caprimulgus atripennis	R	LC
Andaman Nightjar	Caprimulgus andamanicus	r	LC
Indian Nightjar	Caprimulgus asiaticus	R	LC
Savanna Nightjar	Caprimulgus affinis	r	LC
Hemiprocnidae (Treeswift)			
Crested Treeswift	Hemiprocne coronata	R	LC
Apodidae (Swifts)			
Glossy Swiftlet	Collocalia esculenta	R	LC

Common English Name	Scientific Name	Status	IUCN
Indian Swiftlet	Collocalia unicolor	R	LC
Himalayan Swiftlet	Collocalia brevirostris	R	LC
Edible-nest Swiftlet	Collocalia fuciphaga	R	LC
White-rumped Spinetail	Zoonavena sylvatica	R	LC
White-throated Needletail	Hirundapus caudacutus	s	LC
Silver-backed Needletail	Hirundapus cochinchinensis	r	LC
Brown-backed Needletail	Hirundapus giganteus	R	LC
Asian Palm Swift	Cypsiurus balasiensis	R	LC
Alpine Swift	Tachymarptis melba	r	LC
Common Swift	Apus apus	s	LC
Pallid Swift	Apus pallidus	w	LC
Fork-tailed Swift	Apus pacificus	r	LC
Dark-rumped Swift	Apus acuticauda	r	VU
Little Swift	Apus affinis	R	LC
House Swift	Apus a. nipalensis	r	LC
Trogonidae (Trogons)			
Malabar Trogon	Harpactes fasciatus	r	LC
Red-headed Trogon	Harpactes erythrocephalus	r	LC
Ward's Trogon	Harpactes wardi	r	NT
Coraciidae (Rollers)			
Eurasian Roller	Coracias garrulus	rp	NT
Indian Roller	Coracias benghalensis	R	LC
Dollarbird	Eurystomus orientalis	r	LC
Alcedinidae (Kingfishers)			
Blyth's Kingfisher	Alcedo hercules	r	NT
Common Kingfisher	Alcedo atthis	R	LC
Blue-eared Kingfisher	Alcedo meninting	r	LC
Oriental Dwarf Kingfisher	Ceyx erithaca	r	LC
Brown-winged Kingfisher	Pelargopsis amauroptera	r	NT
Stork-billed Kingfisher	Pelargopsis capensis	R	LC
Ruddy Kingfisher	Halcyon coromanda	r	LC
White-throated Kingfisher	Halcyon smyrnensis	R	LC
Black-capped Kingfisher	Halcyon pileata	R	LC
Collared Kingfisher	Todiramphus chloris	r	LC
Crested Kingfisher	Megaceryle lugubris	R	LC
Pied Kingfisher	Ceryle rudis	R	LC
Meropidae (Bee-eaters)			
Blue-bearded Bee-eater	Nyctyornis athertoni	r	LC
Green Bee-eater	Merops orientalis	R	LC
Blue-cheeked Bee-eater	Merops persicus	PS	LC
Blue-tailed Bee-eater	Merops philippinus	R	LC
European Bee-eater	Merops apiaster	sP	LC
Chestnut-headed Bee-eater	Merops leschenaulti	R	LC
Upupidae (Hoopoe)			
Common Hoopoe	Upupa epops	RW	LC
Bucerotidae (Hornbills)			
Malabar Grey Hornbill	Ocyceros griseus	r	LC
Indian Grey Hornbill	Ocyceros birostris	R	LC
Sri Lanka Grey Hornbill	Ocyceros gingalensis	r	LC
Malabar Pied Hornbill	Anthracoceros coronatus	r	NT
Oriental Pied Hornbill	Anthracoceros albirostris	r	LC
Great Hornbill	Buceros bicornis	R	NT
Brown Hornbill	Anorrhinus tickelli	r	NT
Rufous-necked Hornbill	Aceros nipalensis	r	VU
Wreathed Hornbill	Rhyticeros undulatus	r	LC
Narcondam Hornbill	Rhyticeros narcondami	r	EN
Megalaimidae (Barbets)			
Great Barbet	Megalaima virens	R	LC
Brown-headed Barbet	Megalaima zeylanica	R	LC
Lineated Barbet	Megalaima lineata	R	LC
White-cheeked Barbet	Megalaima viridis	R	LC
Yellow-fronted Barbet	Megalaima flavifrons	r	LC
Golden-throated Barbet	Megalaima franklinii	r	LC
Blue-throated Barbet	Megalaima asiatica	R	LC
Blue-eared Barbet	Megalaima australis	r	LC
Malabar Barbet	Megalaima malabarica	r	LC
Crimson-fronted Barbet	Megalaima rubricapillus	r	LC
Coppersmith Barbet	Megalaima haemacephala	R	LC
Indicatoridae (Honeyguide)			
Yellow-rumped Honeyguide	Indicator xanthonotus	r	NT
Picidae (Woodpeckers)			
Eurasian Wryneck	Jynx torquilla	sw	LC
Speckled Piculet	Picumnus innominatus	r	LC
White-browed Piculet	Sasia ochracea	r	LC

Common English Name	Scientific Name	Status	IUCN
Rufous Woodpecker	Micropternus brachyurus	R	LC
White-bellied Woodpecker	Dryocopus javensis	r	LC
Andaman Woodpecker	Dryocopus hodgei	i	NT
Pale-headed Woodpecker	Gecinulus grantia	r	LC
Bay Woodpecker	Blythipicus pyrrhotis	r	LC
Heart-spotted Woodpecker	Hemicircus canente	r	LC
Great Slaty Woodpecker	Mulleripicus pulverulentus	r	VU
Brown-capped Pygmy Woodpecker	Dendrocopos nanus	R	LC
Grey-capped Pygmy Woodpecker	Dendrocopos canicapillus	R	LC
Brown-fronted Woodpecker	Dendrocopos auriceps	R	LC
Fulvous-breasted Woodpecker	Dendrocopos macei	R	LC
Spot-breasted Woodpecker	Dendrocopos m. analis	r	LC
Stripe-breasted Woodpecker	Dendrocopos atratus	i	LC
Yellow-crowned Woodpecker	Dendrocopos mahrattensis	R	LC
Rufous-bellied Woodpecker	Dendrocopos hyperythrus	r	LC
Crimson-breasted Woodpecker	Dendrocopos cathpharius	r	LC
Darjeeling Woodpecker	Dendrocopos darjellensis	R	LC
Great Spotted Woodpecker	Dendrocopos major	R	LC
Himalayan Woodpecker	Dendrocopos himalayensis	R	LC
Sind Woodpecker	Dendrocopos assimilis	r	LC
Lesser Yellownape	Picus chlorolophus	R	LC
Greater Yellownape	Picus flavinucha	R	LC
Streak-breasted Woodpecker	Picus viridanus	r	LC
Streak-throated Woodpecker	Picus xanthopygaeus	R	LC
Scaly-bellied Woodpecker	Picus squamatus	R	LC
Grey-headed Woodpecker	Picus canus	R	LC
Himalayan Goldenback	Dinopium shorii	R	LC
Common Goldenback	Dinopium javanense	R	LC
Lesser Goldenback	Dinopium benghalense	R	LC
Greater Goldenback	Chrysocolaptes lucidus	R	LC
Crimson-backed Goldenback	Chrysocolaptes l. stricklandi	r	LC
White-naped Woodpecker	Chrysocolaptes festivus	r	LC
Eurylaimidae (Broadbills)			
Silver-breasted Broadbill	Serilophus lunatus	r	LC
Long-tailed Broadbill	Psarisomus dalhousiae	r	LC
Pittidae (Pittas)			
Blue-naped Pitta	Pitta nipalensis	r	LC
Blue Pitta	Pitta cyanea	r	LC
Hooded Pitta	Pitta sordida	r	LC
Indian Pitta	Pitta brachyura	R	LC
Mangrove Pitta	Pitta megarhyncha	r	NT
Tephrodornithidae (Woodshrikes)			
Large Woodshrike	Tephrodornis virgatus	r	LC
Malabar Woodshrike	Tephrodornis v. sylvicola	r	LC
Common Woodshrike	Tephrodornis pondicerianus	R	LC
Sri Lanka Woodshrike	Tephrodornis p. affinis	r	LC
Bar-winged Flycatcher-shrike	Hemipus picatus	R	LC
Artamidae (Woodswallows)			
Ashy Woodswallow	Artamus fuscus	R	LC
White-breasted Woodswallow	Artamus leucorynchus	r	LC
Aegithinidae (Ioras)			
Common Iora	Aegithina tiphia	R	LC
Marshall's Iora	Aegithina nigrolutea	r	LC
Campephagidae (Cuckooshrikes, Minivets)			
Large Cuckooshrike	Coracina macei	r	LC
Bar-bellied Cuckooshrike	Coracina striata	r	LC
Black-winged Cuckooshrike	Coracina melaschistos	r	LC
Black-headed Cuckooshrike	Coracina melanoptera	r	LC
Pied Triller	Lalage nigra	r	LC
Rosy Minivet	Pericrocotus roseus	rw	LC
Swinhoe's Minivet	Pericrocotus cantonensis	?	LC
Ashy Minivet	Pericrocotus divaricatus	w	LC
Small Minivet	Pericrocotus cinnamomeus	R	LC
White-bellied Minivet	Pericrocotus erythropygius	r	LC
Grey-chinned Minivet	Pericrocotus solaris	r	LC
Long-tailed Minivet	Pericrocotus ethologus	R	LC
Short-billed Minivet	Pericrocotus brevirostris	r	LC
Orange Minivet	Pericrocotus flammeus	R	LC
Scarlet Minivet	Pericrocotus f. speciosus	R	LC
Pachycephalidae (Whistler)			
Mangrove Whistler	Pachycephala cinerea	i	LC
Laniidae (Shrikes)			
Red-backed Shrike	Lanius collurio	p	LC
Isabelline Shrike	Lanius isabellinus	W	LC

Common English Name	Scientific Name	Status	IUCN
Brown Shrike	Lanius cristatus	W	LC
Red-tailed Shrike	Lanius phoenicuroides	p	LC
Burmese Shrike	Lanius collurioides	p	LC
Bay-backed Shrike	Lanius vittatus	R	LC
Long-tailed Shrike	Lanius schach	R	LC
Grey-backed Shrike	Lanius tephronotus	rW	LC
Lesser Grey Shrike	Lanius minor	V	LC
Great Grey Shrike	Lanius excubitor	V	LC
Southern Grey Shrike	Lanius meridionalis	R	LC
Steppe Grey Shrike	Lanius m. pallidirostris	r	LC
Oriolidae (Orioles)			
Eurasian Golden Oriole	Oriolus oriolus	V	LC
Indian Golden Oriole	Oriolus o. kundoo	R	LC
Black-naped Oriole	Oriolus chinensis	rw	LC
Slender-billed Oriole	Oriolus tenuirostris	r	LC
Black-hooded Oriole	Oriolus xanthornus	R	LC
Maroon Oriole	Oriolus traillii	r	LC
Dicruridae (Drongos)			
Black Drongo	Dicrurus macrocercus	R	LC
Ashy Drongo	Dicrurus leucophaeus	R	LC
White-bellied Drongo	Dicrurus caerulescens	r	LC
Crow-billed Drongo	Dicrurus annectans	r	LC
Bronzed Drongo	Dicrurus aeneus	r	LC
Lesser Racket-tailed Drongo	Dicrurus remifer	r	LC
Greater Racket-tailed Drongo	Dicrurus paradiseus	r	LC
Sri Lanka Drongo	Dicrurus p. lophorinus	r	LC
Spangled Drongo	Dicrurus hottentottus	R	LC
Andaman Drongo	Dicrurus andamanensis	r	NT
Rhipiduridae (Fantails)			
White-throated Fantail	Rhipidura albicollis	R	LC
White-spotted Fantail	Rhipidura a. albogularis	R	LC
White-browed Fantail	Rhipidura aureola	R	LC
Monarchidae (Monarch flycatchers)			
Black-naped Monarch	Hypothymis azurea	r	LC
Asian Paradise-flycatcher	Terpsiphone paradisi	R	LC
Corvidae (Jays, Magpies, and Crows)			
Eurasian Jay	Garrulus glandarius	R	LC
Black-headed Jay	Garrulus lanceolatus	R	LC
Yellow-billed Blue Magpie	Urocissa flavirostris	R	LC
Red-billed Blue Magpie	Urocissa erythrorhyncha	R	LC
Sri Lanka Blue Magpie	Urocissa ornata	r	VU
Common Green Magpie	Cissa chinensis	r	LC
Rufous Treepie	Dendrocitta vagabunda	R	LC
Grey Treepie	Dendrocitta formosae	R	LC
White-bellied Treepie	Dendrocitta leucogastra	r	LC
Collared Treepie	Dendrocitta frontalis	r	LC
Andaman Treepie	Dendrocitta bayleyi	R	NT
Eurasian Magpie	Pica pica	r	LC
Groundpecker	Pseudopodoces humilis	r	LC
Spotted Nutcracker	Nucifraga caryocatactes	r	LC
Large Spotted Nutcracker	Nucifraga c. multipunctata	r	LC
Red-billed Chough	Pyrrhocorax pyrrhocorax	R	LC
Alpine Chough	Pyrrhocorax graculus	R	LC
Eurasian Jackdaw	Corvus monedula	rw	LC
House Crow	Corvus splendens	R	LC
Rook	Corvus frugilegus	w	LC
Carrion Crow	Corvus corone	rw	LC
Hooded Crow	Corvus c. cornix	w	LC
Large-billed Crow	Corvus macrorhynchos	R	LC
Eastern Jungle Crow	Corvus m. levillantii	R	LC
Indian Jungle Crow	Corvus m. culminatus	R	LC
Brown-necked Raven	Corvus ruficollis	r	LC
Northern Raven	Corvus corax	r	LC
Punjab Raven	Corvus c. subcorax	r	LC
Bombycillidae (Waxwing)			
Bohemian Waxwing	Bombycilla garrulus	V	LC
Hypocoliidae (Hypocolius)			
Grey Hypocolius	Hypocolius ampelinus	w	LC
Stenostiridae			
Yellow-bellied Fantail	Chelidorhynx hypoxantha	R	LC
Grey-headed Canary Flycatcher	Culicicapa ceylonensis	R	LC
Paridae (Tits)			
White-crowned Penduline Tit	Remiz coronatus	r	LC

Common English Name	Scientific Name	Status	IUCN
Fire-capped Tit	Cephalopyrus flammiceps	r	LC
Rufous-naped Tit	Periparus rufonuchalis	r	LC
Rufous-vented Tit	Periparus rubidiventris	r	LC
Coal Tit	Parus ater	r	LC
Grey-crested Tit	Lophophanes dichrous	r	LC
Great Tit	Parus major	R	LC
Green-backed Tit	Parus monticolus	R	LC
White-naped Tit	Parus nuchalis	r	VU
Black-lored Tit	Parus xanthogenys	r	LC
Indian Yellow Tit	Parus x. aplonotus	r	LC
Yellow-cheeked Tit	Parus spilonotus	r	LC
Yellow-browed Tit	Sylviparus modestus	r	LC
Sultan Tit	Melanochlora sultanea	R	LC
White-cheeked Tit	Aegithalos leucogenys	r	LC
Black-throated Tit	Aegithalos concinnus	R	LC
White-throated Tit	Aegithalos niveogularis	r	LC
Rufous-fronted Tit	Aegithalos iouschistos	r	LC
Alaudidae (Larks)			
Singing Bushlark	Mirafra cantillans	r	LC
Indian Bushlark	Mirafra erythroptera	R	LC
Bengal Bushlark	Mirafra assamica	R	LC
Jerdon's Bushlark	Mirafra affinis	R	LC
Black-crowned Sparrow Lark	Eremopterix nigriceps	r	LC
Ashy-crowned Sparrow Lark	Eremopterix grisea	R	LC
Rufous-tailed Lark	Ammomanes phoenicura	R	LC
Desert Lark	Ammomanes deserti	r	LC
Bar-tailed Lark	Ammomanes cinctura	r	LC
Greater Hoopoe Lark	Alaemon alaudipes	r	LC
Bimaculated Lark	Melanocorypha bimaculata	w	LC
Tibetan Lark	Melanocorypha maxima	r	LC
Greater Short-toed Lark	Calandrella brachydactyla	W	LC
Hume's Short-toed Lark	Calandrella acutirostris	rw	LC
Asian Short-toed Lark	Calandrella cheleensis	V	LC
Sand Lark	Calandrella raytal	R	LC
Crested Lark	Galerida cristata	R	LC
Malabar Lark	Galerida malabarica	r	LC
Sykes's Lark	Galerida deva	r	LC
Eurasian Skylark	Alauda arvensis	w	LC
Oriental Skylark	Alauda gulgula	R	LC
Horned Lark	Eremophila alpestris	r	LC
Pycnonotidae (Bulbuls)			
Crested Finchbill	Spizixos canifrons	r	LC
Striated Bulbul	Pycnonotus striatus	R	LC
Grey-headed Bulbul	Pycnonotus priocephalus	r	NT
Black-headed Bulbul	Pycnonotus atriceps	r	LC
Andaman Bulbul	Pycnonotus a. fuscoflavescens	r	LC
Black-capped Bulbul	Pycnonotus melanicterus	r	LC
Black-crested Bulbul	Pycnonotus m. flaviventris	R	LC
Flame-throated Bulbul	Pycnonotus m. gularis	R	LC
Red-whiskered Bulbul	Pycnonotus jocosus	R	LC
White-eared Bulbul	Pycnonotus leucotis	R	LC
Himalayan Bulbul	Pycnonotus leucogenys	R	LC
Red-vented Bulbul	Pycnonotus cafer	R	LC
Yellow-throated Bulbul	Pycnonotus xantholaemus	r	VU
Yellow-eared Bulbul	Pycnonotus penicillatus	r	NT
Flavescent Bulbul	Pycnonotus flavescens	r	LC
White-browed Bulbul	Pycnonotus luteolus	R	LC
White-throated Bulbul	Alophoixus flaveolus	r	LC
Olive Bulbul	Iole virescens	r	LC
Yellow-browed Bulbul	Acritillas indica	r	LC
Ashy Bulbul	Hemixos flavala	r	LC
Mountain Bulbul	Ixos mcclellandii	r	LC
Black Bulbul	Hypsipetes leucocephalus	R	LC
Square-tailed Bulbul	Hypsipetes l. ganeesa	R	LC
Nicobar Bulbul	Ixos nicobariensis	r	NT
Hirundinidae (Swallows and Martins)			
Pale Martin	Riparia diluta	pr	LC
Sand Martin	Riparia riparia	r	LC
Plain Martin	Riparia paludicola	R	LC
Eurasian Crag Martin	Ptyonoprogne rupestris	r	LC
Rock Martin	Ptyonoprogne fuligula	V	LC
Dusky Crag Martin	Ptyonoprogne concolor	R	LC
Barn Swallow	Hirundo rustica	RW	LC
Pacific Swallow	Hirundo tahitica	r	LC

Common English Name	Scientific Name	Status	IUCN
Hill Swallow	Hirundo t. domicola	r	LC
Wire-tailed Swallow	Hirundo smithii	R	LC
Red-rumped Swallow	Cecropis daurica	R	LC
Sri Lanka Swallow	Cecropis d. hyperythra	r	LC
Striated Swallow	Cecropis striolata	r	LC
Streak-throated Swallow	Petrochelidon fluvicola	R	LC
Common House Martin	Delichon urbicum	sp	LC
Asian House Martin	Delichon dasypus	r	LC
Nepal House Martin	Delichon nipalense	r	LC
Cisticolidae (Cisticolas, Prinias and Tailorbirds)			
Zitting Cisticola	Cisticola juncidis	R	LC
Bright-headed Cisticola	Cisticola exilis	r	LC
Rufous-vented Prinia	Prinia burnesii	r	NT
Swamp Prinia	Prinia cinerascens	r	NT
Striated Prinia	Prinia crinigera	R	LC
Black-throated Prinia	Prinia atrogularis	r	LC
Hill Prinia	Prinia a. superciliaris	r	LC
Grey-crowned Prinia	Prinia cinereocapilla	r	VU
Rufous-fronted Prinia	Prinia buchanani	R	LC
Rufescent Prinia	Prinia rufescens	r	LC
Grey-breasted Prinia	Prinia hodgsonii	R	LC
Graceful Prinia	Prinia gracilis	R	LC
Jungle Prinia	Prinia sylvatica	R	LC
Yellow-bellied Prinia	Prinia flaviventris	R	LC
Ashy Prinia	Prinia socialis	R	LC
Plain Prinia	Prinia inornata	R	LC
Common Tailorbird	Orthotomus sutorius	R	LC
Dark-necked Tailorbird	Orthotomus atrogularis	w	LC
Cettiidae (Bush Warblers)			
Pale-footed Bush Warbler	Cettia pallidipes	r	LC
Brownish-flanked Bush Warbler	Cettia fortipes	r	LC
Aberrant Bush Warbler	Cettia flavolivacea	r	LC
Hume's Bush Warbler	Cettia brunnescens	r	LC
Grey-sided Bush Warbler	Cettia brunnifrons	r	LC
Cetti's Bush Warbler	Cettia cetti	w	LC
Chestnut-crowned Bush Warbler	Cettia major	r	LC
Chestnut-headed Tesia	Oligura castaneocoronata	r	LC
Slaty-bellied Tesia	Tesia olivea	r	LC
Grey-bellied Tesia	Tesia cyaniventer	r	LC
Yellow-bellied Warbler	Abroscopus superciliaris	r	LC
Rufous-faced Warbler	Abroscopus albogularis	r	LC
Black-faced Warbler	Abroscopus schisticeps	r	LC
Mountain Tailorbird	Phyllergates cuculatus	w	LC
Broad-billed Warbler	Tickellia hodgsoni	r	LC
Acrocephalidae (Reed Warblers)			
Black-browed Reed Warbler	Acrocephalus bistrigiceps	w	LC
Moustached Warbler	Acrocephalus melanopogon	rw	LC
Sedge Warbler	Acrocephalus schoenobaenus	?	LC
Paddyfield Warbler	Acrocephalus agricola	w	LC
Large-billed Reed Warbler	Acrocephalus orinus	r	LC
Blunt-winged Warbler	Acrocephalus concinens	r	LC
Blyth's Reed Warbler	Acrocephalus dumetorum	W	LC
Great Reed Warbler	Acrocephalus arundinaceus	w	LC
Oriental Reed Warbler	Acrocephalus orientalis	w	LC
Clamorous Reed Warbler	Acrocephalus stentoreus	R	LC
Thick-billed Warbler	Phragamaticula aedon	w	LC
Booted Warbler	Iduna caligata	rw	LC
Sykes's Warbler	Iduna rama	rw	LC
Locustellidae (Locustella Warblers and Grassbirds)			
Spotted Bush Warbler	Bradypterus thoracicus	r	LC
West Himalayan Bush Warbler	Bradypterus kashmirensis	s	LC
Long-billed Bush Warbler	Bradypterus major	Nr	NT
Chinese Bush Warbler	Bradypterus tacsanowskius	V	LC
Brown Bush Warbler	Bradypterus luteoventris	r	LC
Russet Bush Warbler	Bradypterus mandelli	r	LC
Sri Lanka Bush Warbler	Elaphrornis palliseri	r	NT
Lanceolated Warbler	Locustella lanceolata	w	LC
Grasshopper Warbler	Locustella naevia	w	LC
Rusty-rumped Warbler	Locustella certhiola	w	LC
Striated Grassbird	Megalurus palustris	R	LC
Bristled Grassbird	Chaetornis striata	r	VU
Rufous-rumped Grassbird	Graminicola bengalensis	r	NT
Broad-tailed Grassbird	Schoenicola platyurus	r	LC
Sylviidae (Sylvia Warblers)			

Common English Name	Scientific Name	Status	IUCN
Garden Warbler	*Sylvia borin*	V	LC
Common Whitethroat	*Sylvia communis*	p	LC
Lesser Whitethroat	*Sylvia curruca*	W	LC
Desert Whitethroat	*Sylvia c. minula*	w	LC
Hume's Whitethroat	*Sylvia althaea*	W	LC
Asian Desert Warbler	*Sylvia nana*	w	LC
Barred Warbler	*Sylvia nisoria*	V	LC
Orphean Warbler	*Sylvia hortensis*	rW	LC
Phylloscopidae (Leaf Warblers)			
Common Chiffchaff	*Phylloscopus collybita*	sW	LC
Mountain Chiffchaff	*Phylloscopus sindianus*	Rw	LC
Plain Leaf Warbler	*Phylloscopus neglectus*	w	LC
Dusky Warbler	*Phylloscopus fuscatus*	w	LC
Smoky Warbler	*Phylloscopus fuligiventer*	sw	LC
Tickell's Leaf Warbler	*Phylloscopus affinis*	sW	LC
Buff-throated Warbler	*Phylloscopus subaffinis*	V	LC
Sulphur-bellied Warbler	*Phylloscopus griseolus*	sw	LC
Radde's Warbler	*Phylloscopus schwarzi*	V	LC
Buff-barred Warbler	*Phylloscopus pulcher*	r	LC
Ashy-throated Warbler	*Phylloscopus maculipennis*	r	LC
Lemon-rumped Warbler	*Phylloscopus chloronotus*	rW	LC
Brooks's Leaf Warbler	*Phylloscopus subviridis*	w	LC
Yellow-browed Warbler	*Phylloscopus inornatus*	rW	LC
Hume's Warbler	*Phylloscopus humei*	W	LC
Mandelli's Leaf Warbler	*Phylloscopus h. mandellii*	wp	LC
Arctic Warbler	*Phylloscopus borealis*	V	LC
Greenish Warbler	*Phylloscopus trochiloides*	rW	LC
Green Warbler	*Phylloscopus t. nitidus*	wp	LC
Pale-legged Leaf Warbler	*Phylloscopus tenellipes*	V	LC
Large-billed Leaf Warbler	*Phylloscopus magnirostris*	rw	LC
Tytler's Leaf Warbler	*Phylloscopus tytleri*	r	NT
Western Crowned Warbler	*Phylloscopus occipitalis*	r	LC
Eastern Crowned Warbler	*Phylloscopus coronatus*	V	LC
Blyth's Leaf Warbler	*Phylloscopus reguloides*	r	LC
Yellow-vented Warbler	*Phylloscopus cantator*	r	LC
Green-crowned Warbler	*Seicercus burkii*	R	LC
Grey-crowned Warbler	*Seicercus tephrocephalus*	R	LC
Grey-hooded Warbler	*Seicercus xanthoschistos*	R	LC
White-spectacled Warbler	*Seicercus affinis*	r	LC
Whistler's Warbler	*Seicercus whistleri*	r	LC
Grey-cheeked Warbler	*Seicercus poliogenys*	r	LC
Chestnut-crowned Warbler	*Seicercus castaniceps*	r	LC
White-browed Tit Warbler	*Leptopoecile sophiae*	r	LC
Timaliidae (Babblers)			
Ashy-headed Laughingthrush	*Garrulax cinereifrons*	r	VU
White-throated Laughingthrush	*Garrulax albogularis*	R	LC
White-crested Laughingthrush	*Garrulax leucolophus*	R	LC
Lesser Necklaced Laughingthrush	*Garrulax monileger*	r	LC
Greater Necklaced Laughingthrush	*Garrulax pectoralis*	r	LC
Striated Laughingthrush	*Garrulax striatus*	r	LC
Rufous-necked Laughingthrush	*Garrulax ruficollis*	r	LC
Chestnut-backed Laughingthrush	*Garrulax nuchalis*	r	NT
Yellow-throated Laughingthrush	*Garrulax galbanus*	i	LC
Wynaad Laughingthrush	*Garrulax delesserti*	r	LC
Rufous-vented Laughingthrush	*Garrulax gularis*	r	LC
Moustached Laughingthrush	*Garrulax cineraceus*	r	LC
Rufous-chinned Laughingthrush	*Garrulax rufogularis*	r	LC
Spotted Laughingthrush	*Garrulax ocellatus*	r	LC
Grey-sided Laughingthrush	*Garrulax caerulatus*	r	LC
Spot-breasted Laughingthrush	*Garrulax merulinus*	r	LC
White-browed Laughingthrush	*Garrulax sannio*	r	LC
Black-chinned Laughingthrush	*Garrulax cachinnans*	r	EN
Kerala Laughingthrush	*Garrulax fairbanki*	r	LC
Grey-breasted Laughingthrush	*Garrulax jerdoni*	r	NT
Streaked Laughingthrush	*Garrulax lineatus*	R	LC
Bhutan Laughingthrush	*Garrulax imbricatus*	R	LC
Striped Laughingthrush	*Garrulax virgatus*	r	LC
Brown-capped Laughingthrush	*Garrulax austeni*	r	LC
Blue-winged Laughingthrush	*Garrulax squamatus*	r	LC
Scaly Laughingthrush	*Garrulax subunicolor*	r	LC
Variegated Laughingthrush	*Garrulax variegatus*	i	LC
Brown-cheeked Laughingthrush	*Garrulax henrici*	?	LC
Black-faced Laughingthrush	*Garrulax affinis*	r	LC
Chestnut-crowned Laughingthrush	*Garrulax erythrocephalus*	r	LC

Common English Name	Scientific Name	Status	IUCN
Assam Laughingthrush	Garrulax chrysopterus	r	LC
Red-faced Liocichla	Liocichla phoenicea	r	LC
Bugun Liocichla	Liocichla bugunorum	r	VU
Abbott's Babbler	Malacocincla abbotti	r	LC
Buff-breasted Babbler	Pellorneum tickelli	r	LC
Spot-throated Babbler	Pellorneum albiventre	r	LC
Marsh Babbler	Pellorneum palustre	r	VU
Puff-throated Babbler	Pellorneum ruficeps	R	LC
Brown-capped Babbler	Pellorneum fuscocapillus	r	LC
Large Scimitar Babbler	Pomatorhinus hypoleucos	r	LC
Spot-breasted Scimitar Babbler	Pomatorhinus erythrocnemis	r	LC
Rusty-cheeked Scimitar Babbler	Pomatorhinus erythrogenys	r	LC
White-browed Scimitar Babbler	Pomatorhinus schisticeps	r	LC
Indian Scimitar Babbler	Pomatorhinus horsfieldii	r	LC
Sri Lanka Scimitar Babbler	Pomatorhinus melanurus	r	LC
Streak-breasted Scimitar Babbler	Pomatorhinus ruficollis	r	LC
Red-billed Scimitar Babbler	Pomatorhinus ochraciceps	r	LC
Coral-billed Scimitar Babbler	Pomatorhinus ferruginosus	r	LC
Slender-billed Scimitar Babbler	Xiphirhynchus superciliaris	r	LC
Long-billed Wren Babbler	Rimator malacoptilus	r	LC
Streaked Wren Babbler	Napothera brevicaudata	r	LC
Eyebrowed Wren Babbler	Napothera epilepidota	r	LC
Scaly-breasted Wren Babbler	Pnoepyga albiventer	r	LC
Nepal Wren Babbler	Pnoepyga immaculata	r	LC
Pygmy Wren Babbler	Pnoepyga pusilla	r	LC
Rufous-throated Wren Babbler	Spelaeornis caudatus	r	NT
Rusty-throated Wren Babbler	Spelaeornis badeigularis	r	VU
Bar-winged Wren Babbler	Spelaeornis troglodytoides	r	LC
Spotted Wren Babbler	Elachura formosa	r	LC
Long-tailed Wren Babbler	Spelaeornis chocolatinus	r	NT
Grey-bellied Wren Babbler	Spelaeornis reptatus	r	NT
Chin Hills Wren Babbler	Spelaeornis oatesi	r	NT
Tawny-breasted Wren Babbler	Spelaeornis longicaudatus	r	VU
Himalayan Wedge-billed Wren Babbler	Sphenocichla humei	r	NT
Manipur Wedge-billed Wren Babbler	Sphenocichla roberti	r	NT
Rufous-fronted Babbler	Stachyridopsis rufifrons	r	LC
Rufous-capped Babbler	Stachyridopsis ruficeps	r	LC
Black-chinned Babbler	Stachyridopsis pyrrhops	R	LC
Golden Babbler	Stachyridopsis chrysaea	r	LC
Grey-throated Babbler	Stachyris nigriceps	r	LC
Snowy-throated Babbler	Stachyris oglei	r	VU
Tawny-bellied Babbler	Dumetia hyperythra	R	LC
Dark-fronted Babbler	Rhopocichla atriceps	r	LC
Pin-striped Tit Babbler	Macronous gularis	R	LC
Chestnut-capped Babbler	Timalia pileata	R	LC
Yellow-eyed Babbler	Chrysomma sinense	R	LC
Jerdon's Babbler	Chrysomma altirostre	r	VU
Common Babbler	Turdoides caudata	R	LC
Striated Babbler	Turdoides earlei	R	LC
Slender-billed Babbler	Turdoides longirostris	r	VU
Large Grey Babbler	Turdoides malcolmi	R	LC
Rufous Babbler	Turdoides subrufa	r	LC
Spiny Babbler	Turdoides nipalensis	r	LC
Jungle Babbler	Turdoides striata	R	LC
Orange-billed Babbler	Turdoides rufescens	R	LC
Yellow-billed Babbler	Turdoides affinis	R	LC
Chinese Babax	Babax lanceolatus	V	LC
Silver-eared Mesia	Mesia argentauris	r	LC
Red-billed Leiothrix	Leiothrix lutea	r	LC
Himalayan Cutia	Cutia nipalensis	r	LC
Black-headed Shrike Babbler	Pteruthius rufiventer	r	LC
White-browed Shrike Babbler	Pteruthius flaviscapis	r	LC
Green Shrike Babbler	Pteruthius xanthochlorus	r	LC
Black-eared Shrike Babbler	Pteruthius melanotis	r	LC
Chestnut-fronted Shrike Babbler	Pteruthius aenobarbus	r	LC
White-hooded Babbler	Gampsorhynchus rufulus	r	LC
Rusty-fronted Barwing	Actinodura egertoni	r	LC
Hoary-throated Barwing	Actinodura nipalensis	r	LC
Streak-throated Barwing	Actinodura waldeni	r	LC
Blue-winged Siva	Siva cyanouroptera	r	LC
Bar-throated Siva	Siva strigula	r	LC
Red-tailed Minla	Minla ignotincta	r	LC
Golden-breasted Fulvetta	Lioparus chrysotis	r	LC
Yellow-throated Fulvetta	Pseudominla cinerea	r	LC

Common English Name	Scientific Name	Status	IUCN
Rufous-winged Fulvetta	Pseudominla castaneceps	r	LC
White-browed Fulvetta	Fulvetta vinipectus	r	LC
Manipur Fulvetta	Fulvetta manipurensis	r	LC
Brown-throated Fulvetta	Fulvetta ludlowi	r	LC
Rufous-throated Fulvetta	Schoeniparus rufogularis	r	LC
Rusty-capped Fulvetta	Schoeniparus dubius	r	LC
Brown-cheeked Fulvetta	Alcippe poioicephala	R	LC
Nepal Fulvetta	Alcippe nipalensis	r	LC
Rufous-backed Sibia	Leioptila annectans	r	LC
Rufous Sibia	Malacias capistratus	R	LC
Grey Sibia	Malacias gracilis	r	LC
Beautiful Sibia	Malacias pulchellus	r	LC
Long-tailed Sibia	Heterophasia picaoides	r	LC
Striated Yuhina	Staphida castaniceps	r	LC
White-naped Yuhina	Yuhina bakeri	r	LC
Whiskered Yuhina	Yuhina flavicollis	R	LC
Stripe-throated Yuhina	Yuhina gularis	R	LC
Rufous-vented Yuhina	Yuhina occipitalis	r	LC
Black-chinned Yuhina	Yuhina nigrimenta	R	LC
White-bellied Erpornis	Erpornis zantholeuca	r	LC
Fire-tailed Myzornis	Myzornis pyrrhoura	r	LC
Paradoxornithidae (Parrotbills)			
Great Parrotbill	Conostoma aemodium	r	LC
Brown Parrotbill	Cholornis unicolor	r	LC
Grey-headed Parrotbill	Psittiparus gularis	r	LC
Black-breasted Parrotbill	Paradoxornis flavirostris	r	VU
Spot-breasted Parrotbill	Paradoxornis guttaticollis	r	LC
Fulvous Parrotbill	Suthora fulvifrons	r	LC
Black-throated Parrotbill	Suthora nipalensis	r	LC
Lesser Rufous-headed Parrotbill	Chleuasicus atrosuperciliaris	r	LC
Greater Rufous-headed Parrotbill	Psittiparus ruficeps	r	LC
Zosteropidae (White-eyes)			
Oriental White-eye	Zosterops palpebrosus	R	LC
Sri Lanka White-eye	Zosterops ceylonensis	R	LC
Irenidae (Fairy Bluebird)			
Asian Fairy Bluebird	Irena puella	R	LC
Regulidae (Goldcrest)			
Goldcrest	Regulus regulus	r	LC
Troglodytidae (Wren)			
Eurasian Wren	Troglodytes troglodytes	r	LC
Sittidae (Nuthatches)			
Chestnut-vented Nuthatch	Sitta nagaensis	r	LC
Kashmir Nuthatch	Sitta cashmirensis	r	LC
Indian Nuthatch	Sitta castanea	R	LC
Chestnut-bellied Nuthatch	Sitta c. cinnamoventris	R	LC
White-tailed Nuthatch	Sitta himalayensis	r	LC
White-cheeked Nuthatch	Sitta leucopsis	r	LC
Velvet-fronted Nuthatch	Sitta frontalis	R	LC
Beautiful Nuthatch	Sitta formosa	r	VU
Trichodromidae (Wallcreeper)			
Wallcreeper	Tichodroma muraria	rw	LC
Certhiidae (Treecreepers)			
Hodgson's Treecreeper	Certhia hodgsoni	R	LC
Bar-tailed Treecreeper	Certhia himalayana	R	LC
Rusty-flanked Treecreeper	Certhia nipalensis	r	LC
Brown-throated Treecreeper	Certhia discolor	r	LC
Hume's Treecreeper	Certhia manipurensis	r	LC
Spotted Creeper	Salpornis spilonotus	r	LC
Sturnidae (Starlings and Mynas)			
Asian Glossy Starling	Aplonis panayensis	r	LC
Spot-winged Starling	Saroglossa spiloptera	r	LC
White-faced Starling	Sturnornis albofrontatus	V	VU
Chestnut-tailed Starling	Sturnia malabarica	R	LC
Blyth's Starling	Sturnia m. blythii	R	LC
White-headed Starling	Sturnia erythropygia	r	LC
Brahminy Starling	Sturnia pagodarum	R	LC
Daurian Starling	Agropsar sturninus	V	LC
Chestnut-cheeked Starling	Agropsar philippensis	V	LC
White-shouldered Starling	Sturnia sinensis	V	LC
Rosy Starling	Pastor roseus	WP	LC
Common Starling	Sturnus vulgaris	wp	LC
Asian Pied Starling	Gracupica contra	R	LC
Great Myna	Acridotheres grandis	V	LC
Common Myna	Acridotheres tristis	R	LC

Common English Name	Scientific Name	Status	IUCN
Bank Myna	*Acridotheres ginginianus*	R	LC
Jungle Myna	*Acridotheres fuscus*	R	LC
Collared Myna	*Acridotheres albocinctus*	r	LC
Golden-crested Myna	*Ampeliceps coronatus*	r	LC
Sri Lanka Hill Myna	*Gracula ptilogeny*	r	NT
Common Hill Myna	*Gracula religiosa*	r	LC
Lesser Hill Myna	*Gracula r. indica*	r	LC
Turdinae (Thrushes)			
Malabar Whistling Thrush	*Myophonus horsfieldii*	R	LC
Blue Whistling Thrush	*Myophonus caeruleus*	R	LC
Sri Lanka Whistling Thrush	*Myophonus blighi*	r	EN
Pied Thrush	*Zoothera wardii*	sp	LC
Orange-headed Thrush	*Zoothera citrina*	R	LC
Siberian Thrush	*Zoothera sibirica*	w	LC
Spot-winged Thrush	*Zoothera spiloptera*	r	NT
Plain-backed Thrush	*Zoothera mollissima*	r	LC
Long-tailed Thrush	*Zoothera dixoni*	r	LC
Scaly Thrush	*Zoothera dauma*	r	LC
Nilgiri Thrush	*Zoothera d. neilgherriensis*	r	LC
Sri Lanka Thrush	*Zoothera imbricata*	r	LC
Long-billed Thrush	*Zoothera monticola*	r	LC
Dark-sided Thrush	*Zoothera marginata*	r	LC
Tickell's Thrush	*Turdus unicolor*	R	LC
Black-breasted Thrush	*Turdus dissimilis*	w	LC
White-collared Blackbird	*Turdus albocinctus*	r	LC
Grey-winged Blackbird	*Turdus boulboul*	r	LC
Common Blackbird	*Turdus merula*	r	LC
Tibetan Blackbird	*Turdus m. maximus*	r	LC
Indian Blackbird	*Turdus m. simillimus*	r	LC
Chestnut Thrush	*Turdus rubrocanus*	r	LC
Kessler's Thrush	*Turdus kessleri*	V	LC
Grey-sided Thrush	*Turdus feae*	w	VU
Eyebrowed Thrush	*Turdus obscurus*	w	LC
Red-throated Thrush	*Turdus ruficollis*	w	LC
Black-throated Thrush	*Turdus atrogularis*	w	LC
Dusky Thrush	*Turdus eunomus*	w	LC
Fieldfare	*Turdus pilaris*	V	LC
Song Thrush	*Turdus philomelos*	V	LC
Mistle Thrush	*Turdus viscivorus*	r	LC
Purple Cochoa	*Cochoa purpurea*	r	LC
Green Cochoa	*Cochoa viridis*	r	LC
Muscicapidae (Flycatchers)			
Gould's Shortwing	*Heterozenicus stellatus*	r	LC
Rusty-bellied Shortwing	*Brachypteryx hyperythra*	r	NT
Lesser Shortwing	*Brachypteryx leucophrys*	r	LC
White-browed Shortwing	*Brachypteryx montana*	r	LC
Nicobar Jungle Flycatcher	*Rhinomyias nicobaricus*	wr	LC
Spotted Flycatcher	*Muscicapa striata*	p	LC
Dark-sided Flycatcher	*Muscicapa sibirica*	r	LC
Asian Brown Flycatcher	*Muscicapa dauurica*	rw	LC
Rusty-tailed Flycatcher	*Muscicapa ruficauda*	r	LC
Brown-breasted Flycatcher	*Muscicapa muttui*	r	LC
Ferruginous Flycatcher	*Muscicapa ferruginea*	r	LC
Yellow-rumped Flycatcher	*Ficedula zanthopygia*	V	LC
Slaty-backed Flycatcher	*Ficedula hodgsonii*	r	LC
Rufous-gorgeted Flycatcher	*Ficedula strophiata*	r	LC
Red-breasted Flycatcher	*Ficedula parva*	W	LC
Taiga Flycatcher	*Ficedula albicilla*	W	LC
Kashmir Flycatcher	*Ficedula subrubra*	r	VU
White-gorgeted Flycatcher	*Anthipes monileger*	r	LC
Snowy-browed Flycatcher	*Ficedula hyperythra*	r	LC
Little Pied Flycatcher	*Ficedula westermanni*	r	LC
Ultramarine Flycatcher	*Ficedula superciliaris*	r	LC
Slaty-blue Flycatcher	*Ficedula tricolor*	r	LC
Sapphire Flycatcher	*Ficedula sapphira*	r	LC
Black-and-orange Flycatcher	*Ficedula nigrorufa*	r	NT
Verditer Flycatcher	*Eumyias thalassinus*	R	LC
Nilgiri Flycatcher	*Eumyias albicaudatus*	r	NT
Dull-blue Flycatcher	*Eumyias sordidus*	r	NT
Large Niltava	*Niltava grandis*	r	LC
Small Niltava	*Niltava macgrigoriae*	r	LC
Rufous-bellied Niltava	*Niltava sundara*	r	LC
Vivid Niltava	*Niltava vivida*	r	LC
White-tailed Flycatcher	*Cyornis concretus*	r	LC

Common English Name	Scientific Name	Status	IUCN
White-bellied Blue Flycatcher	Cyornis pallipes	r	LC
Pale-chinned Flycatcher	Cyornis poliogenys	r	LC
Pale Blue Flycatcher	Cyornis unicolor	r	LC
Blue-throated Blue Flycatcher	Cyornis rubeculoides	r	LC
Hill Blue Flycatcher	Cyornis banyumas	r	LC
Large Blue Flycatcher	Cyornis magnirostris	s	LC
Tickell's Blue Flycatcher	Cyornis tickelliae	R	LC
Pygmy Blue Flycatcher	Muscicapella hodgsoni	r	LC
European Robin	Erithacus rubecula	V	LC
Common Nightingale	Luscinia megarhynchos	V	LC
Siberian Rubythroat	Luscinia calliope	w	LC
White-tailed Rubythroat	Luscinia pectoralis	rW	LC
Bluethroat	Luscinia svecica	sW	LC
Firethroat	Luscinia pectardens	V	NT
Indian Blue Robin	Luscinia brunnea	r	LC
Siberian Blue Robin	Luscinia cyane	V	LC
Himalayan Bluetail	Tarsiger c. rufilatus	r	LC
Golden Bush Robin	Tarsiger chrysaeus	r	LC
White-browed Bush Robin	Tarsiger indicus	r	LC
Rufous-breasted Bush Robin	Tarsiger hyperythrus	r	LC
Rufous-tailed Scrub Robin	Cercotrichas galactotes	p	LC
Oriental Magpie Robin	Copsychus saularis	R	LC
White-rumped Shama	Copsychus malabaricus	R	LC
Andaman Shama	Copsychus m. albiventris	r	LC
Indian Robin	Saxicoloides fulicatus	R	LC
Eversmann's Redstart	Phoenicurus erythronotus	w	LC
Blue-capped Redstart	Phoenicurus coeruleocephala	r	LC
Black Redstart	Phoenicurus ochruros	rW	LC
Common Redstart	Phoenicurus phoenicurus	p	LC
Hodgson's Redstart	Phoenicurus hodgsoni	w	LC
White-throated Redstart	Phoenicurus schisticeps	r	LC
Daurian Redstart	Phoenicurus auroreus	rw	LC
Guldenstadt's Redstart	Phoenicurus erythrogastrus	r	LC
Blue-fronted Redstart	Phoenicurus frontalis	r	LC
White-capped Water Redstart	Chaimarrornis leucocephalus	r	LC
Plumbeous Water Redstart	Rhyacornis fuliginosa	r	LC
White-bellied Redstart	Hodgsonius phaenicuroides	r	LC
Nilgiri Blue Robin	Myiomela major	r	EN
White-bellied Blue Robin	Myiomela albiventris	r	EN
White-tailed Robin	Myiomela leucura	r	LC
Blue-fronted Robin	Cinclidium frontale	r	LC
Grandala	Grandala coelicolor	r	LC
Little Forktail	Enicurus scouleri	r	LC
Black-backed Forktail	Enicurus immaculatus	r	LC
Slaty-backed Forktail	Enicurus schistaceus	r	LC
White-crowned Forktail	Enicurus leschenaulti	r	LC
Spotted Forktail	Enicurus maculatus	r	LC
Stoliczka's Bushchat	Saxicola macrorhynchus	r	VU
Hodgson's Bushchat	Saxicola insignus	w	VU
Common Stonechat	Saxicola torquatus	R	LC
White-tailed Stonechat	Saxicola leucurus	r	LC
Pied Bushchat	Saxicola caprata	R	LC
Jerdon's Bushchat	Saxicola jerdoni	r	LC
Grey Bushchat	Saxicola ferreus	R	LC
Hume's Wheatear	Oenanthe albonigra	r	LC
Northern Wheatear	Oenanthe oenanthe	P	LC
Variable Wheatear	Oenanthe picata	rw	LC
Finsch's Wheatear	Oenanthe finschii	w	LC
Hooded Wheatear	Oenanthe monacha	r	LC
Pied Wheatear	Oenanthe pleschanka	rw	LC
Red-tailed Wheatear	Oenanthe chrysopygia	rw	LC
Desert Wheatear	Oenanthe deserti	rw	LC
Isabelline Wheatear	Oenanthe isabellina	rw	LC
Brown Rock Chat	Cercomela fusca	R	LC
Rufous-tailed Rock Thrush	Monticola saxatilis	sp	LC
Blue-capped Rock Thrush	Monticola cinclorhynchus	R	LC
Chestnut-bellied Rock Thrush	Monticola rufiventris	r	LC
Blue Rock Thrush	Monticola solitarius	rW	LC
Cinclidae (Dippers)			
White-throated Dipper	Cinclus cinclus	r	LC
Brown Dipper	Cinclus pallasii	R	LC
Chloropseidae (Leafbirds)			
Blue-winged Leafbird	Chloropsis cochinchinensis	R	LC
Jerdon's Leafbird	Chloropsis jerdoni	R	LC

Common English Name	Scientific Name	Status	IUCN
Golden-fronted Leafbird	*Chloropsis aurifrons*	R	LC
Orange-bellied Leafbird	*Chloropsis hardwickii*	r	LC
Dicaeidae (Flowerpeckers)			
Thick-billed Flowerpecker	*Dicaeum agile*	R	LC
Yellow-vented Flowerpecker	*Dicaeum chrysorrheum*	r	LC
Yellow-bellied Flowerpecker	*Dicaeum melanoxanthum*	r	LC
Legge's Flowerpecker	*Dicaeum vincens*	r	NT
Orange-bellied Flowerpecker	*Dicaeum trigonostigma*	r	LC
Pale-billed Flowerpecker	*Dicaeum erythrorhynchos*	R	LC
Nilgiri Flowerpecker	*Dicaeum concolor*	r	LC
Plain Flowerpecker	*Dicaeum c. minullum*	r	LC
Andaman Flowerpecker	*Dicaeum c. virescens*	r	LC
Fire-breasted Flowerpecker	*Dicaeum ignipectus*	r	LC
Scarlet-backed Flowerpecker	*Dicaeum cruentatum*	r	LC
Nectariniidae (Sunbirds and Spiderhunters)			
Ruby-cheeked Sunbird	*Chalcoparia singalensis*	r	LC
Purple-rumped Sunbird	*Leptocoma zeylonica*	R	LC
Crimson-backed Sunbird	*Leptocoma minima*	r	LC
Purple-throated Sunbird	*Leptocoma sperata*	r	LC
Olive-backed Sunbird	*Cinnyris jugularis*	r	LC
Purple Sunbird	*Cinnyris asiaticus*	R	LC
Loten's Sunbird	*Cinnyris lotenia*	R	LC
Mrs Gould's Sunbird	*Aethopyga gouldiae*	r	LC
Green-tailed Sunbird	*Aethopyga nipalensis*	r	LC
Black-throated Sunbird	*Aethopyga saturata*	r	LC
Crimson Sunbird	*Aethopyga siparaja*	R	LC
Vigor's Sunbird	*Aethopyga s. vigorsii*	R	LC
Fire-tailed Sunbird	*Aethopyga ignicauda*	r	LC
Little Spiderhunter	*Arachnothera longirostra*	r	LC
Streaked Spiderhunter	*Arachnothera magna*	r	LC
Passeridae (Sparrows and Snowfinches)			
House Sparrow	*Passer domesticus*	R	LC
Spanish Sparrow	*Passer hispaniolensis*	w	LC
Sind Sparrow	*Passer pyrrhonotus*	r	LC
Russet Sparrow	*Passer rutilans*	R	LC
Eurasian Tree Sparrow	*Passer montanus*	R	LC
Rock Sparrow	*Petronia petronia*	w	LC
Chestnut-shouldered Petronia	*Gymnoris xanthocollis*	R	LC
Tibetan Snowfinch	*Montifringilla adamsi*	r	LC
White-rumped Snowfinch	*Onychostruthus taczanowskii*	r	LC
Rufous-necked Snowfinch	*Pyrgilauda ruficollis*	w	LC
Blanford's Snowfinch	*Pyrgilauda blanfordi*	w	LC
Ploceidae (Weavers)			
Black-breasted Weaver	*Ploceus benghalensis*	R	LC
Streaked Weaver	*Ploceus manyar*	R	LC
Baya Weaver	*Ploceus philippinus*	R	LC
Finn's Weaver	*Ploceus megarhynchus*	r	VU's
Estrildidae (Avadavats and Munias)			
Red Avadavat	*Amandava amandava*	R	LC
Green Avadavat	*Amandava formosa*	r	VU
Indian Silverbill	*Euodice malabarica*	R	LC
White-rumped Munia	*Lonchura striata*	R	LC
Black-throated Munia	*Lonchura kelaarti*	r	LC
Scaly-breasted Munia	*Lonchura punctulata*	R	LC
Black-headed Munia	*Lonchura malacca*	R	LC
Chestnut Munia	*Lonchura m. atricapilla*	r	LC
Java Sparrow	*Lonchura oryzivora*	I	LC
Prunellidae (Accentors)			
Alpine Accentor	*Prunella collaris*	r	LC
Altai Accentor	*Prunella himalayana*	w	LC
Robin Accentor	*Prunella rubeculoides*	R	LC
Rufous-breasted Accentor	*Prunella strophiata*	r	LC
Brown Accentor	*Prunella fulvescens*	r	LC
Black-throated Accentor	*Prunella atrogularis*	w	LC
Maroon-backed Accentor	*Prunella immaculata*	r	LC
Motacillidae (Wagtails and Pipits)			
Forest Wagtail	*Dendronanthus indicus*	rW	LC
White Wagtail	*Motacilla alba*	rW	LC
White-browed Wagtail	*Motacilla maderaspatensis*	R	LC
Citrine Wagtail	*Motacilla citreola*	rW	LC
Yellow Wagtail	*Motacilla flava*	W	LC
Grey Wagtail	*Motacilla cinerea*	rW	LC
Richard's Pipit	*Anthus richardi*	W	LC
Paddyfield Pipit	*Anthus rufulus*	R	LC
Tawny Pipit	*Anthus campestris*	W	LC

Common English Name	Scientific Name	Status	IUCN
Blyth's Pipit	Anthus godlewskii	w	LC
Long-billed Pipit	Anthus similis	Rw	LC
Tree Pipit	Anthus trivialis	rW	LC
Olive-backed Pipit	Anthus hodgsoni	RW	LC
Red-throated Pipit	Anthus cervinus	p	LC
Rosy Pipit	Anthus roseatus	r	LC
Water Pipit	Anthus spinoletta	w	LC
Buff-bellied Pipit	Anthus rubescens	w	LC
Upland Pipit	Anthus sylvanus	r	LC
Nilgiri Pipit	Anthus nilghiriensis	r	VU
Fringillidae (Finches)			
Common Chaffinch	Fringilla coelebs	V	LC
Brambling	Fringilla montifringilla	V	LC
Tibetan Serin	Serinus tibetanus	r	LC
Red-fronted Serin	Serinus pusillus	r	LC
Yellow-breasted Greenfinch	Carduelis spinoides	R	LC
Black-headed Greenfinch	Carduelis ambigua	V	LC
Eurasian Siskin	Carduelis spinus	V	LC
European Goldfinch	Carduelis carduelis	r	LC
Twite	Carduelis flavirostris	r	LC
Eurasian Linnet	Carduelis cannabina	W	LC
Plain Mountain Finch	Leucosticte nemoricola	r	LC
Brandt's Mountain Finch	Leucosticte brandti	r	LC
Sillem's Mountain Finch	Leucosticte sillemi	?	?
Spectacled Finch	Callacanthis burtoni	r	LC
Trumpeter Finch	Bucanetes githagineus	w	LC
Mongolian Finch	Eremopsaltria mongolica	w	LC
Blanford's Rosefinch	Carpodacus rubescens	r	LC
Dark-breasted Rosefinch	Carpodacus nipalensis	r	LC
Common Rosefinch	Carpodacus erythrinus	rW	LC
Beautiful Rosefinch	Carpodacus pulcherrimus	r	LC
Pink-browed Rosefinch	Carpodacus rodochroa	r	LC
Vinaceous Rosefinch	Carpodacus vinaceus	r	LC
Dark-rumped Rosefinch	Carpodacus edwardsii	r	LC
Three-banded Rosefinch	Carpodacus trifasciatus	V	LC
Spot-winged Rosefinch	Carpodacus rodopeplus	r	LC
White-browed Rosefinch	Carpodacus thura	r	LC
Red-mantled Rosefinch	Carpodacus rhodochlamys	r	LC
Streaked Rosefinch	Carpodacus rubicilloides	r	LC
Great Rosefinch	Carpodacus rubicilla	r	LC
Red-fronted Rosefinch	Carpodacus puniceus	r	LC
Crimson-browed Finch	Propyrrhula subhimachala	r	LC
Scarlet Finch	Haematospiza sipahi	r	LC
Red Crossbill	Loxia curvirostra	r	LC
Brown Bullfinch	Pyrrhula nipalensis	r	LC
Orange Bullfinch	Pyrrhula aurantiaca	r	LC
Red-headed Bullfinch	Pyrrhula erythrocephala	r	LC
Grey-headed Bullfinch	Pyrrhula erythaca	r	LC
Hawfinch	Coccothraustes coccothraustes	w	LC
Black-and-yellow Grosbeak	Mycerobas icterioides	r	LC
Collared Grosbeak	Mycerobas affinis	r	LC
Spot-winged Grosbeak	Mycerobas melanozanthos	r	LC
White-winged Grosbeak	Mycerobas carnipes	r	LC
Gold-naped Finch	Pyrrhoplectes epauletta	r	LC
Emberizidae (Buntings)			
Crested Bunting	Melophus lathami	R	LC
Yellowhammer	Emberiza citrinella	W	LC
Pine Bunting	Emberiza leucocephalos	w	LC
Rock Bunting	Emberiza cia	R	LC
Godlewski's Bunting	Emberiza godlewskii	?	LC
Grey-necked Bunting	Emberiza buchanani	W	LC
Ortolan Bunting	Emberiza hortulana	V	LC
White-capped Bunting	Emberiza stewarti	rw	LC
Striolated Bunting	Emberiza striolata	r	LC
Chestnut-eared Bunting	Emberiza fucata	rw	LC
Little Bunting	Emberiza pusilla	w	LC
Rustic Bunting	Emberiza rustica	?	LC
Yellow-breasted Bunting	Emberiza aureola	w	VU
Chestnut Bunting	Emberiza rutila	w	LC
Black-headed Bunting	Emberiza melanocephala	w	LC
Red-headed Bunting	Emberiza bruniceps	w	LC
Black-faced Bunting	Emberiza spodocephala	w	LC
Common Reed Bunting	Emberiza schoeniclus	rw	LC
Corn Bunting	Emberiza calandra	V	LC

▪ FURTHER READING ▪

Ali, S. & Ripley, D. (1964-74) *Handbook of the Birds of India & Pakistan* (Vols. 1-10). Bombay: OUP.

Ali, S. & Ripley, D. (1983) *A Pictorial Guide to the Birds of the Indian Subcontinent*. Bombay: OUP.

Ali, S. (1996) *The Book of Indian Birds*. (12th ed.) New Delhi: BNHS & OUP.

Fleming, R.L. Sr. et al. (1984) *Birds of Nepal* Nepal: Nature Himalayas.

Grewal, B, Pfister, O & Harvey, B. (2002) *A Photographic Guide to the Birds of India,* Singapore: Periplus.

Grimmet, R Inskipp, T., & Inskipp, C. (1998) *Birds of the Indian Subcontinent*. UK: A&C Black.

Grimmet R, Inskipp, T & Roberts, T. (2008) *Birds of Pakistan*, London: Christopher Helm.

Harrison, J (1999) *A Field Guide to The Birds of Sri Lanka*, New Delhi: OUP.

Harvey, B, Devasar, N & Grewal, B. (2006) *Atlas of the Birds of Delhi and Haryana*, New Delhi: Rupa & Co.

Kotagama, S. & Fernando, P. (1994) *A Field Guide to the Birds of Sri Lanka*. Colombo: The Wildlife Heritage Trust of Sri Lanka.

Lainer, H (2004) *Birds of Goa*, Goa: The Other India Bookstore.

Naoroji, R (2006) *Birds of Prey of the Indian Subcontinent*, London: Christopher Helm.

Rasmussen, P & Anderton, J (2005) *Birds of South Asia: The Ripley Guide*. Barcelona: Lynx Editions.

Ripley, D. (1982) *A Synopsis of the Birds of India and Pakistan*. Bombay: BNHS.

Roberts, T.J. (1991-92) *The Birds of Pakistan*. Vols. I-2. Karachi: OUP.

ACKNOWLEDGEMENTS

Bikram Grewal and Garima Bhatia would like to thank the following people for their help in producing this book: Aasheesh Pittie, Abhishek Das, Alka Vaidya, Alpa Seth, Amano Samarpan, Ambika Singh, Amit Thakurta, Anand Arya, Anil Lal, Arka Sarkar, Arpit Deomurari, Arshiya Sethi, Aruna Lal, Atanu Mondal, Atul Jain, Barsha Gogoi, Bhaskar Das, Bishan Monnappa, Biswapriya Rahut, Bittu Sahgal, Caeser Sen, Clement M Francis, Deboshree Gogoi, Devashis Deb, Dhritiman Chatterjee, Dipankar Ghose, Gayathri Naik, Gillian Wright, Gunjan Arora, Gunjan Bhatia, Jainy Kuriakose, Jitendra Bhatia, Keya Khare, Koshy Koshy, Kshounish Sankar Ray, Madhavi Raj, Manjula Mathur, Manoj Sharma, Megh Roy Choudhary, Mohit Aggarwal, Mousumi Dutta, Ned Bertz, Neha Sinha, Nikhat Grewal, Nikhil Devasar, Ninad Kulkarni, Niranjan Sant, Nitin Srinivasamurthy, Pakhi Bhatia, Pintz Gajjar, Pradeep Sachdeva, Prasanna Parab, Purnima Dutta, Radhika Singh, Raj Kamal Phukan, Rajat Bhargava, Rajneesh Suvarna, Ramki Sreenivasan, Rani Mitra, Ranjan K Das, Ranjeet Ranade, Rathika Ramasamy, Ratna Ghosh, Samiha Grewal, Sharad Sridhar, Shashank Dalvi, Shekar Bhatia, Soma Jha, Subhoranjan Sen, Sudha Arora, Sudhir Vyas, Sue Varma, Sujan Chatterjee, Sujata Rangaswamy, Sumit K Sen, Sunitha Rangaswamy, Supriyo Samanta, Surya Prakash, Tanaji Chakravarti, Uma Kandasarma, and Vijay Sethi. We would especially like to thank Alpana Khare and her team of Neeraj Aggarwal, Raghuvir Khare, Diya Kapur, Ajmal Nayab Siddiqui and Raj Kishore Beck.

■ INDEX ■